Mathematical Methods in Dynamic Economics

Also by András Simonovits

CYCLES AND STAGNATION IN SOCIALIST ECONOMIES: A Mathematical Analysis

PLANNING, SHORTAGE AND TRANSFORMATION: Festschrift for J. Kornai (*co-editor with E. Maskin*)

PRICES, GROWTH AND CYCLES: Bródy's Festschrift (*co-editor with A. Steenge*)

Mathematical Methods in Dynamic Economics

András Simonovits
Scientific Adviser
Institute of Economics
Hungarian Academy of Sciences
Budapest

First published in Great Britain 2000 by
MACMILLAN PRESS LTD
Houndmills, Basingstoke, Hampshire RG21 6XS and London
Companies and representatives throughout the world

A catalogue record for this book is available from the British Library.

ISBN 0–333–77818–9

First published in the United States of America 2000 by
ST. MARTIN'S PRESS, INC.,
Scholarly and Reference Division,
175 Fifth Avenue, New York, N.Y. 10010

ISBN 0–312–22940–2

Library of Congress Cataloging-in-Publication Data
Simonovits, András.
Mathematical methods in dynamic economics / András Simonovits.
p.   cm.
Includes bibliographical references and index.
ISBN 0–312–22940–2 (cloth)
1. Economics, Mathematical.   2. Statics and dynamics (Social
sciences)—Mathematical models.   I. Title.
HB135.S547   1999
330'.01'51—dc21   99–42139
                                                            CIP

This book is printed on paper suitable for recycling and made from fully managed and sustained
forest sources.

10   9   8   7   6   5   4   3   2   1
09 08 07 06 05 04 03 02 01 00

Printed and bound in Great Britain by
Antony Rowe Ltd, Chippenham, Wiltshire

CONTENTS

Foreword

Introduction 1

v

FOREWORD

Almost from the beginning of my research career (1970), I have been working on Dynamic Problems. Since 1988, I have been teaching economic dynamics at the Budapest University of Economics (called Karl Marx University of Economics till 1990). After several years of experience, I realized that the students need a relatively short lecture note on the mathematical methods of the dynamic economics. Or turning the idea upside down: I have to present those dynamic economic models where the most important mathematic methods are applicable.

In 1993 the College László Rajk (Budapest University of Economics) accepted my proposal to teach a one-year course on the subject. The financial help of the Hungarian Soros Foundation was also helpful. Since 1993 I have also been giving a revised course to the participants of the Ph.D. Program for Economics at that university.

The critical remarks of the participants of these courses have contributed to the improvement on the original material. In the Spring of 1995 I repeated these lectures at the Department of Economics, University of Linz. For that occasion I translated Part I of the notes into English. Since then the Hungarian version has been published by the KJK (Publisher of Economics and Law), and the present English version is an abridged variant of the former.

other hand, I forsake the sophisticated proofs which use deep results of Functional Analysis, Measure Theory and other higher mathematical fields. As a result, my approach is more demanding than that of Chiang (1984) but less demanding than that of Stokey and Lucas (1989). Only the reader can decide if this trial is successful or not. By choosing among the innumerable dynamic economic models, I wanted to present classic models like Hicks' trade cycle model and Samuelson's price dynamics. On the other hand, I also wished to present fresh material, including some of my own works. I only hope that the reader will accept this mix.

At my permanent affiliation (Institute of Economics, Hungarian Academy of Sciences) as well as at my visiting posts, a good deal of colleagues have influenced me. At the first place I mention János Kornai, who introduced me into the topics *Non-Price Control* and *Macroeconomics of the Shortage Economy*. (Unfortunately, papers dealing with the later topic had no place in the English version.) Our long cooperation is reflected not only in our joint publications but in other, subsequent works: both types will be reported in the book. Cars Hommes and Helena Nusse have taught me the modern chapters of nonlinear dynamics during a joint research (which has also been dropped from the English version). György Molnár was a most valuable co-author in modelling *Overlapping Cohorts*.

I acknowledge the direct and indirect help of (in alphabetical order) András Bródy, Zsuzsa Kapitány, Michael Lovell and Béla Martos in the field *Non-Price Control*; Tamás Bauer, John Burkett, Attila Chikán, László Halpern, Mária Lackó and Attila K. Soós in the field *Macroeconomics of the Shortage Economy*; Mária Augusztinovics and Eduardo Siandra in the field *Overlapping Cohorts*. Special thanks are due to Loránd Ambrus-Lakatos, Gábor Kertesi and Leonard Mirman for arousing my interest for dynamic optimization. Last but not least, I express my thanks to Katalin Balla, Zsolt Darvas, Péter Eső, Viktória Kocsis, Gyula Magyarkúti, Balázs Romhányi, Imre Szabó, Péter Tallos, János Vincze and above all, György Molnár for their useful comments on earlier versions. I owe a special obligation to Michael Landesmann and Ernő Zalai for their moral support. Miklós Simonovits developed a program converting my ugly WORD 5.0 files into the formidable TEX files. Miklós Buzai (JETSET) drew the figures and Katalin Fried joined the text and the figures. Of course, none of the persons mentioned is responsible for any error in the book.

I express my gratitude for the financial and moral support to my institute (IE, HAS), the Hungarian Science Foundation (OTKA T 019696), to the Fulbright Committee and the following foreign institutions in chronological order: CORE (Louvain-la-Neuve, Belgium), University of Modena (Italy), University of Illinois at Urbana-Champaign, Wesleyan University (Middletown, CT), University of Linz (Austria), the Universities of Groningen and Tilburg (both in the Netherlands), Boston University and Central European University (Budapest).

Budapest, April 1999.

INTRODUCTION

In this book, we shall deal with relatively simple dynamic economic models. The presentation of each model is preceded by the discussion of the mathematical prerequisites. More precisely, the subsequent mathematical and economic chapters form pairs: each mathematical chapter (with odd number) prepares the ground for the subsequent economic chapter (with even number). There are exceptions at the end of the book: Appendix A contains some auxiliary material from Linear Algebra and Appendices B and C are devoted to additional economic material. Of course, both mathematical and economic chapters are built on their previous counterparts. This structure separates the mathematical groundwork from economic models, but requires the reader to study mathematical problems without knowing their economic use.

The book does not aim at completeness. The basic objective is to present some important methods and models of economic dynamics. The table of content gives an idea on the content of the book, therefore in this introductory chapter I will try only to underline the main characteristics of the book.

PROBLEMS

In the Introduction we shall first overview the most important problems: statics versus dynamics, discrete or continuous time, optimize or not, control-theoretical framework, stability and viability, linearity versus nonlinearity, deterministic versus stochastic models, consumers with finite or infinite lifespan, aggregate versus disaggregate models,

1

elegance versus relevance. Some important ideas are taken from Kornai and Martos (1981a).

Statics versus dynamics

Traditional mathematical economics is *static*: there are no two variables with different time-periods which enter the same equation. For example, in the basic model of general equilibrium (Arrow and Debreu, 1954) the vectors of demand, supply and price refer to a single time-period. There are *quasi-dynamic* models where variables with different time-periods are connected, but in a trivial way (for example, the equilibrium volume and price paths of the Neumann model, (Neumann, 1938)). The real economy, however, is *dynamic*: variables of different time-periods are related to each other in an essential way (Frisch, 1933).

Discrete or continuous time

The duality of discrete and continuous time plays an outstanding role in both mathematics and economics. The daily and physical concepts of time favor the continuous approach, while the tools of measurement and calculations prefer the discrete method. The mathematical theories of time-processes generally work with continuous time. Their basic tools are the *differential equations*. Time is the independent variable, and the dependent variables and their derivatives enter the same system of equations. For example, in the differential equation of the mathematical pendulum, the acceleration (that is, the second derivative of the position) is approximately proportional to the position where the proportionality factor is negative.

In contrast, numerical analysis inevitably applies discrete time (steps), and relies on *difference equations*. Time is again the independent variable, but now the dependent variables and their differences or their lagged values enter the same system of equations. For example, in case of annual capitalization, capital and its increment are connected by the interest rate.

On the one hand, continuous-time systems are more complex than discrete-time systems, since the existence and the uniqueness of a solution to differential equations are already deep mathematical results. On the other hand, continuous-time systems are less complex than discrete-time systems, since the resulting paths are simpler in the former than in the latter. For example, in the scalar case no cycle can arise, in the planar case chaos is excluded in the former but not in the latter. It is of particular interest that the analysis of continuous-time systems can be approximated by discrete-time systems when the length of the time-period goes to zero. Among the first economic models we find both continuous and discrete-time models. For example, Samuelson (1939a) modelled the accelerator–multiplier interaction with a discrete-time framework, while Samuelson (1941) described the dynamic adjustment of prices by a continuous-time model. Moreover, the first business cycle models used mixed (difference–differential) equations (for example, Frisch, 1933). Since the pioneers appeared in the scene, both approaches have had extended literature. As a consequence, this book also follows both styles. However, it would be inappropriate to work out both frameworks at full length. Depending on the economic content we shall present sometimes one, sometimes the other method. (If I had to choose between them in economic applications, I would select the discrete framework, because it gives a better approximation to economic problems more often than its continuous counterpart.)

Optimize or not?

The first dynamic economic models (the Walrasian price dynamics by Samuelson, the Hicksian trade cycle model and so on) were not based on optimization. It took several decades that Ramsey (1928) was followed by modern dynamic optimization models (for example, Tinbergen, 1960). The models without optimization are presently unpopular, since the decision-makers' behavior is not derived from first principles. It is of interest that in physics it is useful to derive the objective laws of motion from the maximization of certain functions, although nobody thinks that anybody executes this otherwise meaningless optimization.

The 'genuinely modern' authors simply exclude models without optimization from the realm of mathematical economics. For example, commenting on Solow's realistic assumption of time-invariant saving ratio in his famous neoclassical growth model, Azariadis (1993, p. 4) wrote: "Solow made an *ad hoc* assumption – and there are few sins as grave as this for a self-respecting economist."

Some 'moderately modern' books take a balanced view concerning optimization. Blanchard and Fischer (1989, p. 28) argue as follows: "Our neoclassical bent does not extend to thinking that the

only valid macroeconomic models are those explicitly based on maximization. ...[W]e believe that waiting for a model based on first
principles before willing to analyze current events and give policy ad-
vice is a harmful utopia that leaves the real world to the charlatans
rather than to those who recognize the uncertainties of our current
knowledge."

The present author is indifferent whether the behavioral assumptions are stylized facts of real life or derived from optimization under
constraints. (It is ironic that a number of economists use the distinction *descriptive* versus *normative* approaches and clearly prefer the
latter to the former.) As a defense of the nonoptimizing approach,
I mention three factors: (i) Historical interest mentioned above.
(ii) Most dynamic optimization models have only one decision-maker
and it is questionable whom the *representative agent* represents (Kir-
mann, 1992). Anyway, critique on the exaggerated role of optimization in economics is still relevant (Kornai, 1971, Nelson and Winter,
1982 and Anderson et al., 1988). (iii) The insistence on optimization makes the mathematical treatment more difficult than otherwise,
preventing to follow the principle *first things first*.

Correspondingly, the material is divided into two parts: Parts
I and II deal with economic dynamics without and with optimization, respectively. For example, the control of a multisector economy
would be difficult to describe as optimal strategies of decision-makers,
especially a single one. On the other hand, to understand life-cycle
problems of one person or the whole society's consumption path, the
application of intertemporal optimization may be relevant.

Control-theoretic framework

The book often uses the control-theoretic framework. We shall speak
of a *control system* if the variables of a dynamical system are divided into two groups: *state* and *control* variables. We are given
a *state-space equation*, which describes the feasible dynamics of the
state vector as a function of the *control vector*. Perhaps the simplest
control is the *feedback* when the control vector only depends on the
current state vector. Among other advocates of the control theoretical approach, Kornai and Martos (1981a, b) convincingly argue for
the advantages of this approach. As an illustration, we shall choose
a leading example from the book edited by them: the change in the
output stock of a product is equal to output less sales. In the simplest

control by stock signal, the output is a decreasing function of the output stock (see for example, Section 2.2). In the bulk of the economic models, *decentralization* is much emphasized. To generalize the previous example: suppose that the whole economy is controlled by stock signals, then the change in the output stock of a firm is equal to its output less total sales, while its output is still a decreasing function of its own output stock. At the same time, the macrobehavior rules are not decentralized.

Notwithstanding its advantages, the control-theoretic approach is not exclusive in the dynamic economics. For example, in the closed exchange models of Overlapping Generations and Cohorts (Appendices B and C) the potential state variable (the accumulated saving) is fixed at zero, thus control theory is not applicable.

Stability and viability

It occurs very frequently that the path of the dynamical system cannot or need not be explicitly determined. This is not always a serious problem, because we are often interested in the qualitative rather than the quantitative behavior of the system. The methods described below help answer our questions.

As a starting point, let us consider the *fixed points*, that is, the rest points of the system: if a path starts from there, it stays there forever. In natural sciences a fixed point is frequently called an *equilibrium point*, or simply, an *equilibrium*. In economic applications, the notion of equilibrium is often restricted to that of *walrasian* equilibrium (see Section 6.3 below) where the self-regulating market mechanism ensures the equality of demand and supply. Since Keynes (1936) economists have also been speaking of *unemployment equilibrium* and since the 1970's of *nonwalrasian equilibrium*. Kornai (1980) and Kornai and Martos (1981b) prefer the more neutral expression *normal state*. Since in Overlapping Generations and Cohorts models (Appendices B and C) certain nonstationary paths are also called equilibrium path, we shall also use the expression *steady state* for a stationary path.

The following double question arises: does there exist a fixed point, and if yes, is it uniquely determined? We shall see that three answers are possible: zero, one and multiple equilibria exist. A further question is the *stability* of the equilibrium: if the system does not start from an equilibrium, will it converge to it? More precisely: how large is the basin of attraction of initial points from where the system converges to a fixed point? With a little imprecision: if the basin is large, then the system is globally stable; if it is small, then the system is only locally stable.

In the Nature as well as in the Society *cyclic* processes arise rather frequently. For example, the Earth revolves around the Sun, the seasons follow each other, the human heart beats 60–90 per minutes, the growth rate of the economy oscillates more or less regularly. It is of interest that for every system with a cycle, there exists another system where the states of the cycle of the original system are fixed points. Staying with deterministic systems, stable equilibria and cycles are joined by more complex paths to be called *chaotic*. Then the path depends sensitively on the initial conditions, thus no forecast is possible. Weather is the best-known example of chaos, but it is conceivable that exchange rates also behave chaotically.

In all the three cases one may ask: is the system *viable*? For example, the solar system may function for 10–20 billion years, a human being may live for a hundred years, social systems may survive decades, centuries or millennia. If we study a special economic system, viability means that, in addition to the governing equations, the system satisfies certain conditions as well. In economics the viability conditions are most often nonnegativity conditions. For example, the output cannot be negative. In the model of control by stock signal, inventories are nonnegative but below capacity. We still know little on economic viability and often we have to be content with searching for stability.

Linear versus nonlinear systems

In every mathematical investigation linearity is of paramount interest. With some simplification, a model is homogeneously linear if doubling the values of the input variables doubles those of the output variables. For example, the equation of change in inventory is linear: doubled output and sales imply doubled change in inventories. At first sight the following rule of control by stock signal is also linear: every day at most 100 units are produced, but this maximum is diminished by the double of the end stock of previous day. But what happens if 51 units remained yesterday? Shall we produce –2 units? No. Rather a natural floor (zero lower bound) enters and linearity disappears.

The theory of *finite-dimensional linear systems* is complete. We know the general solution of the system, and with its help we can obtain a lot of quantitative and qualitative results. In this case the local behavior around the equilibrium also determines the global behavior. As a result, cyclic behavior prevails only for exceptional values of parameters (and it has knife-edge behavior). Similarly, unstable behavior implies inviability at least in the long run.

The situation is totally different with *nonlinear systems*. Everything goes already for one-variable nonlinear (discrete-time) systems. For example, local stability can coexist with the lack of global stability; stable cyclic behavior can be observed for a large domain of the parameter space; instability may be consistent with long run viability. As discussed above, in addition to regular *fixed points* and *cyclic* paths, irregular *chaotic* paths also appear. Here we have to rely often on computer simulation even in the two-variable case. Both the mathematicians and the economists had been content with the study of linear or approximately linear systems until quite recently. Only the last decades have witnessed the blossoming of global analysis of nonlinear and unstable dynamical systems. Of course, even in the investigation of such complex systems, the linear system remains the starting point. The book deals with both linear (for example, Chapters 1 and 2) and nonlinear systems (for example, Chapters 3 and 4).

Deterministic versus stochastic models

For modern researchers, the need to study both deterministic and stochastic systems is obvious. It suffices to refer to the coexistence of the classical and the quantum mechanics.

The spread of stochastic approach in dynamic economics is mostly connected to the econometric method. Following the statistical approach, at the estimation of equations it is appropriate to assume stochastic errors. It is small wonder that stochastic optimization plays an outstanding role in the control models based on stochastic equations. These issues are dealt with in Chapter 7. It is noteworthy that in modern dynamic economics the mainstream economists assume that the deterministic core of the economy is stable, and the economy is destabilized *only* by the stochastic shocks. I do not share this view. Following the example of a much smaller but still significant group of mathematical economists, I prefer a cycle theory which is based on deterministic nonlinearities and which also lead to complex dynamics. As a result, in this book stochastic models are underemphasized, nonlinearity is overemphasized, at least with respect to the mainstream approach (Chapters 3–8 and Appendices B and C). We only shortly refer here to a simpler application of stochastic methods, namely, the life insurance problems arising from the uncertainty of human lifespan (Section 10.1).

Expectations

One distinctive feature of economic models is that certain variables may depend on the forecast future values (*expectations*) of other variables. For example, the order that a bookseller places this week may depend on the expected sales next week; or my savings in this year may depend on the interest rate expected for next year. In the models of the 1950's and the 1960's naive (or more generally, adaptive) expectations occurred where the expected value depends on past values (and past expectations). For example, our bookseller naively expects that people will buy that same number of books next week as they did this week. Or, to consider another example, mortgage bankers often calculate the monthly payments of their clients on the assumption that the present interest rate prevails for the future.

This assumption has been criticized since Lucas (1976) and it has been replaced more and more by the assumption of *rational expectations*. In this case for a given information set, the decision-maker's expectation is equal to the expected value of the variable according to the model. In particular, in a deterministic framework, the rational expectation is identical to the actual value, that is, we have *perfect foresight*.

We shall consider both types of expectations. We shall see that indeed in some cases the rational expectations are stable while the naive expectations are not (Section 2.3*). However, there are other cases when the opposite is true (Section 4.4*), moreover, rational expectations practically cannot be defined even in an arbitrary small neighborhood of the steady state (Appendix C).

I wish to contribute to the renaissance of naive expectations which may be attractive both empirically (Shiller, 1978; Lovell, 1986 and Chow, 1989) and theoretically (Grandmont and Laroque, 1990, Brock and Hommes, 1997 and Grandmont, 1998).

Consumers with finite or infinite lifespan

According to an assumption made very frequently, the representative agent has an infinite lifespan in the optimization models. This trick is helpful to avoid problems connected with end-conditions and incompleteness of markets.

This assumption obviously contradicts the finiteness of human life, and its replacement changes the original optimality conditions. In the so-called Overlapping Generations models (Appendix B) the lifespan of the decision-makers is shorter than the horizon: everybody lives two time-periods, first as young, second as old. Every time-period the old and the young live together. The young work much more than the old, who in turn live from the savings accumulated in young age. We only mention two surprises: a) Introduction of an unfunded pension system may improve the welfare of every generation if the growth rate is higher than the interest rate (Samuelson, 1958). b) Notwithstanding optimization, cycles or chaos may emerge (Gale, 1973 and Grandmont, 1985).

This model is generalized by the Overlapping Cohorts models (Appendix C): members of numerous cohorts live together, with age-specific survival rates, wages and consumption units. This generalization often undermines the validity of previously mentioned theorems.

Aggregate versus disaggregate models

In the 1950–60s economists took great pain that their models be disaggregated. In the framework of linear systems, matrices enjoyed then much attention. As Samuelson wrote in one of his papers from that period, that was the age of Leontief. By now, this approach has lost much of its popularity and theoretical economists are often satisfied with one- or two-sector models.

The present book returns to the traditions from this point of view, too. We shall discuss multisector models at various chapters. Chapters 2 and 4 devote much space to the analysis of the control of Leontief economy, while Appendix C studies a multicohort model. Correspondingly, Appendix A summarizes the prerequisites from Linear Algebra.

Elegance versus relevance

The book applies analytical as well as numerical methods. On the one hand, I strive at simplicity. I am always pleased if analytical methods can be used. On the other hand, I often insist on realism. I am not ready to sacrifice relevance at the altar of elegance.

Let me mention a single example. In Appendix C I insist on modelling the interaction of a large number of cohorts, since I cannot accept the foregoing simplifications of the Overlapping Generations model.

I do not pretend to be an applied economist, working with large empirical models on a powerful supercomputer. Nevertheless, the reader is advised to use his PC (or even his pocket calculator) to get some experience with quantitative application of dynamic models.

Finally, let me reinforce my arguments with a quote from Arrow and Honkapohja's (1985a, p. 26): "Two general suggestions about methodology of economic theory were made [at the symposium, A.S.]. The first was that, in view of the number of cases in which theoretical results were nonexistent or too difficult to find, reliance should be placed to a greater extent on numerical simulation of models for various values of the parameters. Simulations would also give a greater feel for the sensitivity of results to specific numerical assumptions. Closely related was the urging that, when theorists model a specific area, they should give some indication of the range of parameter values they would regard as basically consistent with the intended applicability of the model."

Critical attitude

It can be surprising that so many critical remarks can be found in a textbook on mathematical economics. Were it not better to save the reader from the troubles and only show the correct solutions? I think this is not the case. On the one hand, in mathematical economics the real troubles stem not so much from wrong mathematical reasoning but inappropriate economic modelling. On the other hand, even the greatest figures of mathematics or physics have committed errors which make the history of sciences really interesting.

APPROACH

Until now I have tried to clarify the problems highlighted in the book. However, the approach of the book might be more important than the content. Therefore I try to explain the main characteristics of the book, too.

Methods and models

Economic models are chosen to *illustrate* the mathematical techniques. Therefore the book lacks economic coherence, and does not serve as a textbook for economics. (The readers, who do not like such an approach, might find many valuable economic books applying dynamic methods, for example, Blatt (1983), Blanchard and Fischer (1989) and Azariadis (1993)). In this respect I follow the example set by Baumol (1954), Lancaster (1968), Gandolfo (1971), Kamien and Schwartz (1981), Chiang (1984), Stokey and Lucas (1989) and Halanay and Samuel (1998), to name only a few of them.

Multitude of methods

The book does *not* specialize on one or two mathematical methods. Readers interested in one or two methods in detail, are advised to consult other books. The main body of Gandolfo (1971) discussed linear dynamical systems without optimization, Kamien and Schwartz (1981) presented many parts and applications of Calculus of Variations and Theory of Optimal Control; Hommes (1991), Medio (1992) and Day (1994) applied chaos theory to economics. In contrast, our aim is similar to that of Chiang (1984) and Stokey and Lucas (1989): to show the richness of methods.

Level of the book

Originally I called the book *intermediate,* but several readers convinced me that it was an understatement. Anyway, the reader is supposed to know thoroughly calculus, linear algebra and basic micro- and macroeconomics. It is also important that he be interested in the ideas if not the details of the proofs. But I do not aim at completeness of proofs. For example, discussing the existence and unicity of the solution to differential equations, I do not explain the deep methods of functional analysis. I confine the proof of the underlying contraction principle to finite-dimensional vector spaces and simply state that its appropriate generalization in abstract spaces is also valid. This approach is more demanding than that of Chiang (1984) but less demanding than that of Stokey and Lucas (1989).

Examples and problems

The book contains more than a hundred examples and problems. They are located at the appropriate places rather than at the end of the Chapters. The reader is advised to solve the problems immediately or in case of a failure, to consult the Solutions at the end of the book.

Teaching and learning requirements

On the basis of experience, I risk the following requirements on the teaching of the course at a university.

We have already touched on the prerequisite and problem solving (see above), thus we turn to the time requirements. Considering a lecture of 90 minutes per week and serious homework, the whole course can be taught and learned in a year (two semesters).

In case of a shorter course (one semester) or joint problem solving session one can select in several ways. We list some possibilities.

- Mathematical methods (odd chapters: 1, 3, 5, 7 and 9 plus Appendix A).
- Economic models (even chapters: 2, 4, 6, 8, 10, plus Appendices B and C).
- Dynamics without optimization (Part I).
- Dynamics with optimization (Part II and Appendices B and C) – with due preparation (Chapters 1, 3 and 5).
- Dynamic models in discrete time (Chapters 1–4 and 7–8 and Appendices B and C).
- Chaos in economics (Chapters 3, 4 and Appendix B) with Chapter 1 as a preparation.

Starred sections, theorems, proofs, examples and problems are difficult and can be skipped.

As blocks of the same book, the various chapters are connected, however, certain repetitions make them relatively independent.

References

References are to be found at the end of the book and their list contains a good deal of sources which might enhance and deepen the reader's knowledge. Translating a Hungarian text into English, I tried to avoid burdening this book with sources only available in Hungarian.

Fortunately enough, in most cases it was possible to find the English
equivalent of the Hungarian sources, although this modification had
some minor repercussions.

NOTATIONS

We follow the tradition until the elements get in conflict with each
other. Theorems, Examples, Problems and formulas are numbered
separately, and independently in each Chapter, with a double index:
the first number refers to the chapter, while the second one refers to
the foregoing item in the given chapter. In the case of Appendices,
letters A, B and C stand in the first place, respectively.

Mathematical notations

Matrices are denoted by upper case Roman letters, vectors by the
corresponding lower case Roman letters: for example, $A = (a_{ij})$ and
$b = (b_j)$. (Bold symbols are used for distinction but they do not refer
to vectors.) The unit matrix is denoted by I. The null-matrix, the
null vector and zero are equally denoted by 0. $i = \sqrt{-1}$, the Euler
number is e, the Ludolph number $3.14\ldots$ is denoted by π. Inequality
between two vectors (say x and y) is denoted by $x \geq y$ or $x \leq y$, or
$x > y$ or $x < y$. Diagonal matrix is denoted by $M = \langle m \rangle$. A dark
square stands for the end of the proof.

The discrete as well as the continuous *time* variable is denoted
by t, the only distinction made is that in the former a subscript,
in the latter an 'independent variable' in brackets represent it: x_t
and $x(t)$, respectively. x' or $\dot{x}$ stands for the time derivative of $x(t)$,
while the $m \times n$ matrix F_x denotes the derivative matrix of function
$F : \mathbf{R}^m \to \mathbf{R}^n$. To save letters, we shall use the same letters for
corresponding variables and parameters in discrete and continuous-
time systems, even if it is a little bit confusing. (For example, the
continuous-time counterpart of the difference equation $x_t = M x_{t-1}$,
that is, $x_t - x_{t-1} = (M - I)x_{t-1}$ would be $\dot{x} = (M - I)x$ rather than
$\dot{x} = Mx$, but we shall use the latter.) Terminal time and cycle-period
will be denoted by T and P, respectively.

As is usual in control theory, the state and control vectors will
be denoted by x and u, respectively.

A steady state (equilibrium, stationary, normal) or optimal value will be distinguished by ° from its ordinary counterpart, * refers to other distinction, ^{T} for transposition and $\bar{x}$ for complex conjugation. $\Re$ and $\Im$ are real and imaginary parts, respectively.

Considering a vector series, for example, $x_{i,t}$, subindex i and t will denote coordinate i and time index t, respectively. Note the difference between x_F and x_{F}: in the former, symbol F is a mathematical quantity; in the latter, F is a name (for feasibility).

Economic notations

As is usual in macroeconomics, Y, C and I stand for output (GDP), consumption and investment, respectively. Relative (per capita or per GDP) values are denoted by y, c and i, respectively. Γ usually stands for GDP's growth factor (or rate), r for interest factor (or rate). $U(\cdot)$ and $u(\cdot)$ represent the total and the per-period utility functions, respectively.

The same symbol may represent different economic variables or parameters in different models, but we try to avoid ambiguity. Tradition made it impossible to eliminate the important ambiguity with u as a control vector (Chapter 7) and a utility function (Chapter 8) and with f as a transition function (Chapter 1), a production function (Chapter 6) and a reward function (Chapter 7).

Scalar expectations are generally denoted as $_t x_\tau$ for the expected state at time-period t for future time-period τ. Simple vector expectations on v_t will be denoted as v_t^{e}.

PART I

DYNAMICS WITHOUT OPTIMIZATION

In this Part we shall consider such mathematical methods and dynamic economic models which simply assume rather than derive the behavior rules. In the Introduction this approach has already been justified.

Chapter 1 discusses *discrete-time systems* with the help of *difference equations* where the 'time indexes' of the left and right hand sides of certain equations are different. First the *basic concepts* are introduced, then *linear* difference equations are analyzed.

Chapter 2 contains three *linear dynamic economic models* which can be studied with the methods of Chapter 1.

Chapter 3 deals with *nonlinear difference equations* where instability does not imply explosion, cycles are not exceptions and small initial errors can lead to great errors.

Chapter 4 lifts the restriction of linearity imposed in Chapter 2. Our *nonlinear dynamic economic models* prove the applicability of the methods developed in Chapter 3.

Chapter 5 analyzes *continuous-time systems* with the help of *differential equations*, where in addition to the functions, their time-derivatives also enter the equations.

Chapter 6 contains the discussion of three *continuous-time economic models*.

1. LINEAR DIFFERENCE EQUATIONS

In this Chapter we shall consider such systems of equations where each variable has a time index $t = 0, 1, 2, \ldots$, and variables of different time-periods are connected in a nontrivial way. To avoid complicated situations, we assume that there is one variable at the left hand side (to be abbreviated as L.H.S) of each equation, its time index is at least as great as any other in the right hand side (to be abbreviated as R.H.S.) and at least in one equation the time index is definitely greater. Such systems of equations are called *difference equations* or *systems of difference equations*, which describe *dynamical systems with discrete time*. The whole book is built on this chapter.

In Section 1.1 we introduce the *basic concepts* of difference equations. In Sections 1.2–1.4 we shall investigate the simplest systems, namely, the *linear* ones. We shall review the properties of *general, planar and control* (linear) systems, respectively. Useful information can be found in Samuelson (1947), Bodewig (1959), Varga (1962), Ralston (1965), Young (1971), Martos (1981) and Appendix A.

1.1 DIFFERENCE EQUATIONS

This Section introduces such basic concepts of the theory of difference equations like fixed point, stability, cycles and control systems.

Systems of first-order difference equations

Let *time* be a discrete variable, denoted by $t = 0, 1, \ldots$. Let x be a real vector of n dimensions: $x \in \mathbf{R}^n$, and $\{f_t(\cdot)\}_{t=1}^{\infty}$ be a series of transformations of a domain $\mathcal{X} \subseteq \mathbf{R}^n$ into itself. Then the system of equations

(1.1) $x_t = f_t(x_{t-1}), \qquad t = 1, 2, \ldots$

is called a *system of first-order difference equations* or simply *first-order difference equation*.

Example 1.1. Geometrical sequence. Let $\{x_t\}$ be a scalar sequence, $x_t = q x_{t-1}$, $t = 1, 2, \ldots$. Then $x_t = q^t x_0$. (Similarly, for arithmetic sequence, $x_t = x_{t-1} + d$, $t = 1, 2, \ldots$)

It is of interest to rewrite (1.1) as follows:

(1.1*) $x_t - x_{t-1} = f_t^*(x_{t-1})$

where $f_t^*(x_{t-1}) = f_t(x_{t-1}) - x_{t-1}$. Indeed, (1.1*) is a *genuine difference* equation, which is a direct analogy of the differential equation (5.1) to be studied later. Nevertheless, we shall retain the simpler form (1.1).

If the initial state x_0 is given, then system (1.1) uniquely determines *path* $x_0, x_1, x_2, \ldots$.

For the sake of simplicity, we shall mostly confine the analysis to systems which are *time-invariant*:

(1.2) $x_t = f(x_{t-1}).$

In economics this assumption frequently implies that the growth trend is already eliminated.

The form (1.2) is coordinate-free, which is very advantageous by its conciseness. However, it is quite common that the vector difference equation is given in coordinates:

(1.2′) $x_{i,t} = f_i(x_{1,t-1}, \ldots, x_{n,t-1}), \qquad i = 1, \ldots, n$

where $f_i : \mathbf{R}^n \to \mathbf{R}$ functions and $\{x_{i,t}\}_{t=0}^{\infty}$ are scalar sequences.

In certain problems the end values rather than the initial values are given. In other applications for certain variables the end values, for the remaining ones the initial values are given (Chapters 7–10).

Every time-variant system can be described as a time-invariant system if variable t is considered as the $(n+1)$th variable: with $x_{0,t} = t$

(1.2*a) $x_{0,t} = x_{0,t-1} + 1,$

(1.2*b) $x_{i,t} = f_i(t, x_{1,t-1}, \ldots, x_{n,t-1}), \qquad i = 1, \ldots, n;$

that is, using notations $X = (t, x)$ and $F(X) = (t, f(X))$, we have $X_t = F(X_{t-1})$. We shall not use this transformation, however, because it is not practical.

Every explicit difference equation, with everywhere defined R.H.S. and given initial condition, has a unique solution and its analysis appears to be quite simple. If there are side-conditions or we have an implicit difference equation (like in Part II and Appendices B and C), then complications arise. Additional problems are caused by rational expectations, when the dynamics depends not only on the past but as well as on the future. Stochastic disturbances also add to the complexity.

Systems, where the time lag between the L.H.S. and the R.H.S. is longer than 1, are called *higher-order* dynamical systems. In a scalar system the maximal lag defines the *order of the system*. It is well-known that higher-order dynamical systems can be reduced to first-order systems, thus (1.1) is not restrictive. Instead of giving a general formula for the reduction, we present a simple example.

Example 1.2. Reduction of second-order equations. Consider the following second-order difference equation: $y_t = g(y_{t-1}, y_{t-2})$. Let $x_{1,t} = y_t$, $x_{2,t} = y_{t-1}$, then $x_{2,t-1} = y_{t-2}$, that is, the resulting first-order system $x_{1,t} = g(x_{1,t-1}, x_{2,t-1})$ and $x_{2,t} = x_{1,t-1}$ is equivalent to the original second-order one.

Problem 1.1. Mathematical pendulum. Consider the discrete-time version of the frictionless mathematical pendulum (Example 5.6 below). Choose the unit of time as one-quarter of *the period* and measure time from the start. Let y_t be the angle of the pendulum to the vertical line at time-period t. Then the difference equation of the pendulum is $y_t = -y_{t-2}$ with initial conditions $y_0 > 0$, $y_1 = 0$. Execute the reduction of Example 1.2.

Fixed point and stability

Fixed points play a distinguished role in dynamical systems.

A point $x^\circ \in \mathcal{X}$ is called a *fixed point* of the time-invariant system f if, starting the system from x°, it stays there:

$$\text{If} \quad x_0 = x^\circ, \quad \text{then} \quad x_t = x^\circ, \quad t = 1, 2, \ldots.$$

By (1.2), x° is then also a fixed point of map f:

$$(1.3) \qquad x^\circ = f(x^\circ).$$

Remark. Recall the comments made in the Introduction that the fixed point is also called equilibrium, stationary point, steady state and so on but these latter expressions, especially of equilibrium may have other meanings as well.

The following example and problem stand for illustrations.

Example 1.3. Fixed points. In Example 1.1 ($x_t = qx_{t-1}$), if $q = 1$, then every point is a fixed point; if $q \neq 1$, then there exists a unique fixed point: $x^\circ = 0$.

Problem 1.2. Transitional equilibrium? Why does the system of Problem 1.1 not stay in $y_0 = 0$? (Why does the pendulum not stop at the lowest point?)

We now turn to the stability of a fixed point.

1. A fixed point x° of (1.2) is called *Lyapunov stable* if for any close enough initial state x_0, the resulting trajectory $\{x_t\}$ exists and stays close forever to x°. In formula: for any positive real ε, there exists a positive real δ_ε such that if $||x_0 - x^\circ|| < \delta_\varepsilon$, then $||x_t - x^\circ|| < \varepsilon$ for every t. (Here $||x||$ stands for the *norm* of vector x, generalizing the ordinary concept of length, see Appendix A.)

2. A Lyapunov stable fixed point x° is called *locally asymptotically stable* if the path $\{x_t\}$ starting from any initial point x_0 close enough to x°, converges to the fixed point.

3. A Lyapunov stable fixed point is called *globally asymptotically stable* if any path generated by almost any initial point x_0 converges to it. (Of course, if other fixed points exist, they should not be considered for convergence, Example 3.3.)

4. A fixed point is Lyapunov or asymptotically *unstable* if it is not Lyapunov or asymptotically stable, respectively.

Remarks. 1. It is obvious that the existence of a fixed point does not imply its stability.

2. In the economic literature the adjective *asymptotic* is generally omitted, while in mathematics its use is obligatory. In economics the adjective *Lyapunov* is always used, while in mathematics it is not necessary.

3. Some mathematical economists (for example, Arrow and Hahn, 1971) accept that in the definition of global stability the attracting fixed point is not unique.

4. Because of its importance, the *saddle-point instability* deserves a special treatment: then certain initial conditions imply stable, others

unstable paths. If every nonstationary path is unstable, we speak of *complete instability.*

The following example shows why it is not enough to demand convergence without Lyapunov stability in the definition of asymptotic stability.

Example 1.4. (Elagdi, 1991, Example 4.4.) Attraction without the Lyapunov property. A planar system is given in polar coordinates: $r_t = \sqrt{r_{t-1}}$, $\theta_t = \sqrt{2\pi\theta_{t-1}}$ where $r > 0$, $0 \le \theta \le 2\pi$. It is easy to see that the fixed point is $(1,0) = (1, 2\pi)$ and all paths converge to it. Indeed, r_t is increasing (decreasing) for $r_t < 1$ ($r_t > 1$) and θ_t increasingly converges to 2π. Nevertheless, for initial states θ_0 arbitrary close to 0, θ_t wanders far enough before approaching 0.

We return to the pendulum.

Problem 1.3. Stability of the pendulum. a) Is the fixed point (0,0) of the pendulum (asymptotically) stable? b) Is it Lyapunov stable?

Cycles

The concept of cycle is more complicated than that of fixed point, but it is still relatively simple.

Let P be an integer larger than 1. A series of vectors $x_1, x_2, \ldots,$ x_P is called a P-period cycle, or simply a *P-cycle* of system f if a path starting from x_1 goes through $x_2, \ldots, x_P$ and returns to x_1. In formula:

$$(1.4) \qquad x_t = f(x_{t-1}), \qquad t = 2, 3, \ldots, P+1, \qquad x_{P+1} = x_1.$$

As a simple consequence, we have $x_{kP+Q} = x_Q$, for $Q = 1, \ldots, P$, $k = 1, 2, \ldots$.

The simplest cycle arises in the mirror and with the pendulum.

Example 1.5. Cycles in mirror. All the paths of Example 1.1 are 2-cycles for $q = -1$.

Problem 1.4. Cycles of pendulum. a) Determine the cycles of the discrete pendulum. b) How could one get a 2-cycle for a pendulum?

Control systems

The notion of *control system* has a central role in engineering as well
as in economic applications. The basic concepts are as follows: the
state vector and the *control vector*, denoted by x and u, respectively.
The time-invariant *state space equation* determines the new state as
a function of the old state and the new control:

$$(1.5) \qquad x_t = g(x_{t-1}, u_t)$$

where g is an $\mathbf{R}^{n+m} \to \mathbf{R}^n$ function.

A system g is called *controllable* if any initial state x^1 can be
transferred to any end state x^2 by a suitable control sequence $u_1, u_2,$
$\ldots, u_T$.

We speak of *feedback control* if the control vector only depends
on the last state vector:

$$(1.6) \qquad u_t = h(x_{t-1}).$$

The resulting equation will be called the *basic equation*:

$$(1.7) \qquad x_t = f(x_{t-1}) = g[x_{t-1}, h(x_{t-1})].$$

In this case, x° provides a stationary control vector u°:

$$(1.8) \qquad x^\circ = f(x^\circ) \quad \text{and} \quad u^\circ = h(x^\circ).$$

Remark. In economic applications the stock-flow distinction
plays an important role. The state variable is stock, referring to the
end-of-period, for example, capital stock at the end of the time-period.
The control variable is flow, referring to the entire time-period, for
example, annual output. A weakness of the discrete-time models is
the ambiguity whether we classify a stock value as the end value of
time $t-1$ or the initial value of time t. We prefer the first choice. At
the second choice, (1.5)–(1.7) is modified as follows:

$$(1.5') \qquad x_{t+1} = g(x_t, u_t),$$
$$(1.6') \qquad u_t = h(x_t),$$
$$(1.7') \qquad x_{t+1} = f(x_t) = g[x_t, h(x_t)].$$

We speak of *stabilization* if the feedback stabilizes the state space
equation, that is, the fixed point of the basic equation is stable.

The following control system is well-known from everyday life.

Example 1.6. a) Control of room's temperature by heating. Let x_t and w_t be the temperature of a room and that of its external neighborhood, at the end of time-period t, respectively and let u_t be the intensity of the heating during interval t. Then the dynamics of the room temperature is $x_t = A(x_{t-1} - w_t) + Bu_t$ where A and B are appropriate constants. b) Control of room's temperature by thermostat. Let x^* be the room's desired temperature and apply the feedback control $u_t = -K(x_t - x^*) + q_t$. Then the equation of the regulated system is $x_t = A(x_{t-1} - w_t) - BKx_t + BKx^* + Bq_t$. In ideal case, with the proper choice of K and q_t, $x_t = x^*$ can be achieved.

We shall speak of a *(totally) decentralized control* if the number of control variables is equal to that of state variables: $m = n$; and the feedback is decomposed into n independent scalar feedback rules:

$$(1.9) \qquad u_{i,t} = h_i(x_{i,t-1}), \qquad i = 1, \ldots, n.$$

We shall only give examples for controllability and decentralized control in Section 1.4 below, after making the necessary preparations.

1.2 GENERAL LINEAR SYSTEMS

As is usual in mathematics, in the theory of difference equations the linear equations are of paramount importance. In the remaining part of this Chapter we shall only consider such systems. (In Chapter 3 we shall return to nonlinear difference equations.) To make the remaining part of the Chapter more transparent, we put some auxiliary information on Linear Algebra into Appendix A.

System (1.2) is called *linear* if function f is linear. In this case $f(\alpha x + \beta y) = \alpha f(x) + \beta f(y)$ for all $x, y \in \mathbf{R}^n$ and all $\alpha, \beta \in \mathbf{R}$, $\alpha + \beta = 1$. Then there exists an n-dimensional transformation M and an n-vector w, such that $f(x) = Mx + w$. (The restriction $\alpha + \beta = 1$ is required by inhomogeneity.)

Inhomogeneous equations

As a result, a time-invariant, inhomogeneous linear difference equation has the following form:

$$(1.10) \qquad x_t = Mx_{t-1} + w, \qquad t = 1, 2, \ldots.$$

(1.10) is a coordinate-free form. As in (1.2′), we also present the system with coordinates:

$$(1.10') \qquad x_{i,t} = \sum_{j=1}^{n} m_{ij} x_{j,t-1} + w_i, \qquad i = 1, \ldots, n$$

where $M = (m_{ij})$ is the matrix of transformation M in a fixed coordinate system and $w = (w_1, \ldots, w_n)^{\mathrm{T}}$. We need not distinguish a transformation from a matrix like we do not differentiate between a vector without coordinates and that of coordinates.

Let us turn to the fixed points. (1.10) yields the following implicit equation for the fixed point:

$$(1.10°) \qquad\qquad x° = M x° + w.$$

Finally, we quote definitions (A.2)–(A.3) of the *characteristic root* λ, the right hand *characteristic vector* s and the *characteristic polynomial* of transformation M:

$$(1.11) \qquad\qquad M s = \lambda s, \quad s \neq 0,$$
$$(1.12) \qquad\qquad P(\lambda) = \det(\lambda I - M).$$

The conditions of existence and uniqueness of a fixed point are obvious:

Theorem 1.1. *The linear system (1.10) has a unique fixed point, given by*

$$(1.13) \qquad\qquad x° = (I - M)^{-1} w,$$

if and only if 1 is not a characteristic root of M.

Problem 1.5. Let α and β be negative scalars. Determine the fixed point of system

$$M = \begin{pmatrix} 0 & \alpha \\ \beta & 0 \end{pmatrix} \quad \text{and} \quad w = \begin{pmatrix} 1 \\ 1 \end{pmatrix}$$

(see Example A.1) and check Theorem 1.1.

Homogeneous equations

In the theory of systems of algebraic linear equations the connection between solutions of homogeneous and inhomogeneous equations is important. A similar relation exists in our case, too.

Introduce the *deviation vector*

$$(1.14) \qquad\qquad x_t^{\mathrm{d}} = x_t - x^{\circ}$$

and subtract (1.10°) from (1.10):

$$(1.15^{\mathrm{d}}) \qquad\qquad x_t^{\mathrm{d}} = M x_{t-1}^{\mathrm{d}}.$$

In words: the deviation vectors satisfy the homogeneous system (1.15^{d}) obtained from the inhomogeneous system (1.10) by deleting the additive term w. By a step-by-step substitution, (1.15^{d}) yields

$$(1.16^{\mathrm{d}}) \qquad x_t^{\mathrm{d}} = M x_{t-1}^{\mathrm{d}} = M^2 x_{t-2}^{\mathrm{d}} = \cdots = M^t x_0^{\mathrm{d}}.$$

Returning to the original variables, $x_t = x^{\circ} + M^t(x_0 - x^{\circ})$ is obtained. Note that this formula could be derived directly, with the help of the generalized formula of the sum of the geometric sequence.

From now on we shall mostly consider the homogeneous system and for compactness, we shall drop the superscript ^{d}. (We also could say that $w = 0$.) For the sake of reference, we repeat $(1.15^{\mathrm{d}}) - (1.16^{\mathrm{d}})$ in this new notation:

$$(1.15) \qquad\qquad x_t = M x_{t-1},$$
$$(1.16) \qquad\qquad x_t = M^t x_0.$$

In general it is not appropriate to determine x_t for every initial state x_0 by (1.16), using powers of a matrix. This is true even if one applies the more economical iteration $M^t x_0 = M(M^{t-1} x_0)$. However, it is known from linear algebra that M^t can be determined rather simply with the help of the characteristic roots and characteristic vectors of M (see the remarks to Theorem A.4).

The significance of the characteristic roots and characteristic vectors of M is due to the following fact: taking the powers of M the former behave as scalars, the latter as invariants. More exactly:

$$(1.17) \qquad\qquad M^t s = \lambda^t s, \qquad t = 0, 1, 2, \ldots.$$

For the sake of simplicity, let us assume that matrix M is *simple*, that is, there exist n linearly independent characteristic vectors, that is, a *characteristic basis*:

$$(1.18) \qquad P(\lambda_j) = 0, \qquad j = 1, 2, \ldots, n;$$

$$(1.19) \qquad M s_j = \lambda_j s_j, \qquad j = 1, 2, \ldots, n.$$

Then any initial state can be uniquely represented as a linear combination of the characteristic vectors:

$$(1.20) \qquad x_0 = \sum_{j=1}^{n} \xi_j s_j.$$

Using (1.16), (1.17), (1.19) and (1.20) yield

$$(1.21) \qquad x_t = \sum_{j=1}^{n} \xi_j \lambda_j^t s_j.$$

We have proved

Theorem 1.2. *If M has a characteristic basis, then with the help of the characteristic vectors of M, the initial state can be represented as (1.20), and also using the characteristic roots of M, the solution can be represented as (1.21).*

Remark. In practical calculations it is sufficient to determine the characteristic roots and looking for unknown vectors v_j in $x_t = \sum_{j=1}^{n} v_j \lambda_j^t$, $t = 1, 2, \ldots$ from the initial conditions.

The following examples and problems illustrate the working of the method in the simplest cases.

Example 1.7. Diagonal matrix. Let us assume that M is a diagonal matrix: $M = \langle m \rangle$ where $m = (m_j)$ is an n-vector. Then the characteristic roots are $\lambda_j = m_j$ and the characteristic vectors are $s_j = e_j$ (the unit vectors), $j = 1, \ldots, n$. Finally, for $x_0 = (x_{1,0}, \ldots, x_{n,0})^{\mathrm{T}}$, $\xi_j = x_{j,0}$, that is, $x_t = \sum_j x_{j,0} \lambda_j^t e_j$.

Note that a) the block-diagonal structure (Appendix A) is a generalization of the diagonal matrix, and b) the introduction of a characteristic basis (if it exists) *diagonalizes* the original transformation.

Problem 1.6. Calculation. Determine the homogeneous solution of Problem 1.5 with the method just outlined.

Problem 1.7. Lower triangular matrix. How can one simplify the solution for a lower triangular matrix where $m_{ij} = 0$ if $j > i$?

Multiple characteristic roots*

What happens if a characteristic root λ belongs to a characteristic vector s with algebraic multiplicity $r > 1$? Theorem A.5 is now helpful. The whole matrix can be decomposed into a block-diagonal structure. It is sufficient to consider a single block, denote it by M rather than M_{QQ} and it can be described in the following form: $M = \lambda I + N$ where $N^r = 0$. Applying the binomial theorem on M^t, only the first r terms survive: $(\lambda I + N)^t = \lambda^t I + t\lambda^{t-1} N + \ldots + C_{t,r-1}\lambda^{t-r+1} N^{r-1}$ where $C_{t,r}$ stands for the binomial coefficient (t, r). By algebraic arrangement, each entry of M^t is λ^{t-r} times a polynomial of t with degree at most $r - 1$. Finally, denote the *principal vectors* by $s_{j-1} = (\lambda I - M)s_j$, $j = r, r - 1, \ldots, 1$ where s_1 is the single genuine characteristic vector and $s_0 = 0$. Then

$$(1.21^*) \qquad\qquad x_t = \sum_{j=0}^{r} \xi_j \lambda^{t-j} t^j s_j.$$

As a short-cut, many mathematical economists assume that all the characteristic roots are different, ensuring the existence of a characteristic basis. Note, however, that this assumption excludes the identity matrix itself.

Arnold (1973) makes the following witty remark. During the 18th century, when Euler and Lagrange struggled with multiple roots, they did not yet know the Jordan normal form of matrices. Their heuristic approach can be illustrated by the following example with $r = 2$: Approximate matrix M with a sequence of matrices $\{M_k\}$, such that each M_k have different characteristic roots $\lambda_{1,k}$ and $\lambda_{2,k}$, thus every approximating matrix has two independent characteristic vectors $s_{1,k}$ and $s_{2,k}$. Instead of the combination $\xi_1 \lambda_{1,k}^t$ and $\xi_2 \lambda_{2,k}^t$, we rely on the combination $\xi_1 \lambda_{1,k}^t$ and $\delta_k = \xi_2(\lambda_{2,k}^t - \lambda_{1,k}^t)/(\lambda_{2,k} - \lambda_{1,k})$. When $k \to \infty$, then $\lambda_{1,k} \to \lambda$ and $\lambda_{2,k} \to \lambda$, $\delta_k \to \xi_2 t \lambda_1^{t-1}$. Thus $\xi_1 \lambda^t$ is accompanied by $\xi_2 t \lambda^{t-1}$ in (1.21^*).

Complex characteristic roots

Equation (1.21) can be directly used if all the characteristic roots of M are real. What to do, however, if some roots are complex?

Since the entries of matrix M are real, the coefficients of $P(\lambda)$ are also real. According to a well-known elementary algebraic theorem, then each complex characteristic root occurs together with its complex conjugate. Then each characteristic vector s_j and coefficient ξ_j also appear with their complex conjugates.

In fact, because the sum of two homogeneous solutions is also a homogeneous solution, for simple roots, we can assume that $\xi_j = 0$ for $j = 3, 4, \ldots, n$. Let the first complex characteristic root, characteristic vector and coefficient be λ, s and ξ, respectively and the corresponding conjugates be $\bar{\lambda}$, $\bar{s}$ and $\bar{\xi}$:

$$(1.19') \qquad\qquad Ms = \lambda s \quad \text{and} \quad M\bar{s} = \bar{\lambda}\bar{s}.$$

Since $\xi \neq 0$, by normalizing s and $\bar{s}$, we can assume that $\xi = 1$, that is, $\bar{\xi} = 1$. We describe x_0 with the help of the conjugate characteristic vectors:

$$(1.20') \qquad\qquad x_0 = s + \bar{s},$$

and apply the operator M t-times:

$$(1.21') \qquad\qquad x_t = \lambda^t s + \bar{\lambda}^t \bar{s}.$$

Introducing the imaginary root $i = \sqrt{-1}$ and using Moivre's trigonometric formulas $\lambda = |\lambda|(\cos\varphi + i\sin\varphi)$ and $\lambda^t = |\lambda|^t(\cos\varphi t + i\sin\varphi t)$, and substituting into $(1.21')$, yields

$$x_t = |\lambda|^t[(\cos\varphi t + i\sin\varphi t)(\Re s + i\Im s) + (\cos\varphi t - i\sin\varphi t)(\Re s - i\Im s)]$$

where $\Re z$ and $\Im z$ are the *real* and *imaginary parts* of a complex number z, respectively.

Rearranging the terms in brackets, the imaginary terms cancel out, only real terms remain, with multipliers $\cos\varphi t$ and $\sin\varphi t$:

$$(1.22) \qquad\qquad x_t = 2|\lambda|^t[\Re s \cos\varphi t - \Im s \sin\varphi t].$$

We have proved

Theorem 1.3. *If matrix M has a simple complex root λ and a corresponding characteristic vector s, then a corresponding block-solution has the form (1.22).*

Here is the simplest example.

Example 1.8. Rotation by 90° in the plane. Let

$$M = \begin{pmatrix} 0 & -1 \\ 1 & 0 \end{pmatrix}.$$

It is easy to see that $x_t = Mx_{t-1}$ describes an anti-clockwise rotation by 90° in the plane. It is not surprising that $\lambda_{1,2} = \pm i$. With initial conditions $x_{1,0} = 1$ and $x_{2,0} = 0$, the dynamics is $x_{1,t} = \cos(t\pi/2)$ and $x_{2,t} = \sin(t\pi/2)$.

Problem 1.8. Generalization of Example 1.8. Let ρ and φ be two positive reals. Let

$$M = \rho \begin{pmatrix} \cos\varphi & -\sin\varphi \\ \sin\varphi & \cos\varphi \end{pmatrix}.$$

Show that transformation $x_t = Mx_{t-1}$ describes a magnification /contraction by ρ and rotation by φ in the plane. Also demonstrate that $\lambda_{1,2} = \rho[\cos\varphi \pm i\sin\varphi]$, $x_{1,t} = \rho^t \cos\varphi t$ and $x_{2,t} = \rho^t \sin\varphi t$.

Higher-order equations

In discussing Example 1.2 we have already mentioned that there are higher-order difference equations but they can be reduced to first-order equations. We present now such a reduction in case of an n-order scalar homogeneous linear equation with real coefficients $\gamma_1, \ldots, \gamma_n$ and initial conditions $y_0, y_1, \ldots, y_{n-1}$:

$$y_t = \sum_{k=1}^{n} \gamma_k y_{t-k}, \qquad t = n, n+1, \ldots.$$

Introducing the notation $x_{k,t} = y_{t-k}$, with an appropriate matrix M one can describe system $x_t = Mx_{t-1}$. It is simpler, however, to write down the characteristic polynomial of the system: $P(\lambda) = \lambda^n - \sum_{k=1}^{n} \gamma_k \lambda^{n-k}$. Let $\lambda_1, \ldots, \lambda_n$ be the n different roots, then the solution can be expressed with appropriate coefficients $\xi_1, \ldots, \xi_n$ as $y_t = \sum_{k=1}^{n} \xi_k \lambda_k^t$. (It is important to emphasize that in higher-order systems multiple roots have a unique characteristic vector, therefore the Jordan-form applies.)

Here is a classic example.

Example 1.9. Fibonacci numbers (1202). "How many pairs of rabbits will be produced in a year, beginning with a single pair, if every month each pair bears a new pair which becomes productive from the second month on?" (Boyer, 1968, p. 281). It is easy to see that the answer is given by the following recursion: $F_t = F_{t-1} + F_{t-2}$, $F_0 = 1$ and $F_1 = 1$. The characteristic polynomial of the system (or equivalently, the generator function of the series) is $P(\lambda) = \lambda^2 - \lambda - 1$, the characteristic roots are $\lambda_{1,2} = (1 \pm \sqrt{5})/2$. The solution has the form $F_t = \xi_1 \lambda_1^t + \xi_2 \lambda_2^t$. The constants ξ_1 and ξ_2 can be determined from the initial conditions: $F_0 = \xi_1 + \xi_2 = 1$ and $F_1 = \xi_1 \lambda_1 + \xi_2 \lambda_2 = 1$. Hence $\xi_{1,2} = (5 \pm \sqrt{5})/10$.

Stability versus instability

In Section 1.1 we have defined the concepts of local and global stability. We shall need the concept of spectral radius. The *spectral radius* of a square matrix M is the maximum of the moduli of the n characteristic roots, its notation is $\rho(M)$. *Dominant characteristic roots* are characteristic roots with maximal modulus. *Dominant characteristic vectors* are characteristic vectors belonging to dominant roots. For linear systems, the local and the global stability are equivalent and the following theorem can be proved rather easily.

Theorem 1.4. *The discrete-time linear system (1.15) is stable if and only if the spectral radius of M is less than unity:*

$$(1.23) \qquad\qquad \rho(M) < 1.$$

Proof.   a) Assume the existence of a characteristic basis. By (1.21) (see also (1.22)), x_t converges to 0 if and only if all powers of each characteristic root converge to zero, that is, each modulus is less than unity, that is, (1.23) holds. b*) In the general case, (1.21*) gives the same result.   ∎

Remarks.   1. By Theorem 1.4, matrices satisfying (1.23) are called *stable* (in discrete-time framework).

2. If $\rho(M) = 1$, then for simple dominant roots, Lyapunov stability holds; for multiple dominant roots, instability prevails (Problem 1.9 below). For $\rho(M) > 1$, the system is explosive, at least with probability 1 according to initial states.

3. The Neumann series generalizes the traditional formula for the sum of infinite geometric series from scalars in $(-1,1)$ to stable matrices: $(I - M)^{-1} = \sum_{t=0}^{\infty} M^t$.

Problem 1.9. Multiplicity and stability. Compare the stability of solutions of two systems with matrices

$$M_1 = \begin{pmatrix} 0 & -1 \\ -1 & 0 \end{pmatrix} \quad \text{and} \quad M_2 = \begin{pmatrix} 1 & -1 \\ 0 & 1 \end{pmatrix} .$$

Note the difference between M_1 and M_2: M_1 has two simple dominant roots, while M_2 has a single dominant root with multiplicity 2.

To obtain sharper results, we consider *nonnegative irreducible matrices* (Appendix A). By Theorem A.7d, we have a sharpening of Theorem 1.1:

Theorem 1.5. *If matrix M is nonnegative, irreducible and stable, and vector w is positive, then (1.10) has a unique fixed point which is stable and positive.*

We return to general matrices. The speed of convergence at stability is measured by the *dampening factor*, the limit (if it exists) of the reversed ratio of norms of two subsequent deviation states:

$$\Phi = \lim_t \frac{||x_t^d||}{||x_{t+1}^d||}.$$

(If the limit does not exist, we can also take upper limit.)
It is easy to prove

Theorem 1.6. *(Mises.) If the dominant root of M is real, then the dampening factor of iteration (1.10) almost always exists and is given by*

$$\Phi = \frac{1}{\rho(M)}.$$

Proof. If there exists a characteristic basis, then the statement follows from (1.21), under the assumption of $\xi_1 \neq 0$. (The general proof relies on Jordan normal form: (1.21*).) ∎

Assumption $\xi_1 \neq 0$ holds for almost all initial states but if it does not hold, numerical disturbances take care of the problem. The following example shows an illuminating calculation, when $\xi_1 = 0$ (Figure 1.1).

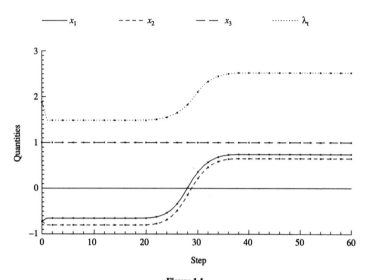

Figure 1.1
Irregular Mises iteration

Example 1.10. Self-correction (Ralston 1965, Example 10.2).

$$M = \begin{pmatrix} 1 & 1 & 0.5 \\ 1 & 1 & 0.25 \\ 0.5 & 0.25 & 1 \end{pmatrix} \quad \text{and} \quad x_0 = \begin{pmatrix} 0.64955116 \\ 0.74822116 \\ 0 \end{pmatrix}.$$

This implies $x_t \approx \xi_2 \lambda_2^t s_2$ for $\xi_1 = 0$ and $t = 2, \dots, 20$, but then round-off errors turn back the system to the right way: $x_t \approx \xi \lambda_1^t s_1$ about from $t = 40$.

Remarks. 1. For complex dominant roots, there is no convergence but $1/\rho(M)$ is still a good indicator on the speed of convergence to or divergence from the fixed point (see Section 1.4 below).

2. It may be surprising that this theorem is used just in the opposite direction in numerical analysis: the dominant root of M is determined by iteration (1.15).

By Theorem A.8b, we have a

Corollary. *If $M \geq 0$ and $M^n > 0$, then the unique dominant root is positive and the dampening factor exists for all positive initial states.*

A special case

The special case $w = 0$ and $\rho(M) = 1$ with a simple unique real dominant root deserves a particular attention. If the dominant root is positive, that is, it is equal to 1, then the dominant (left and right hand) characteristic vectors are fixed points: p_1 and s_1. Of course, we need normalize both of them to obtain a unique solution.

Theorem 1.7. *Consider the singular case of $w = 0$ and $\rho(M) = 1$. Assume that the dominant root $\lambda_1 = 1$, then the corresponding left hand characteristic vector is a fixed point: $p_1^{\mathbf{T}} = p_1^{\mathbf{T}} M$. Consider the hyperplane $p_1^{\mathbf{T}} x = 1$ and normalize s_1 to lie there: $p_1^{\mathbf{T}} s_1 = 1$. Then the members of iteration (1.10) also belong to the hyperplane:*

$$(1.24) \qquad p_1^{\mathbf{T}} x_t = 1, \qquad t = 1, 2, \ldots;$$

and the fixed point s_1 is unique and globally stable within the hyperplane.

Proof. Multiplying (1.16) by p_1: $p_1^{\mathbf{T}} x_t = p_1^{\mathbf{T}} M^t x_0 = p_1^{\mathbf{T}} x_0 = 1$. Multiplying (1.21) by p_1 and applying Theorem A.3: $p_1^{\mathbf{T}} x_t = \sum_j \xi_j \lambda_j^t p_1^{\mathbf{T}} s_j = \xi_1 p_1^{\mathbf{T}} s_1$. By normalization, $\xi_1 = 1$ holds and Theorem 1.6 applies. ∎

Corollary. *If $M \geq 0$ and $M^n > 0$, then the assumptions of uniqueness and simplicity of dominant vectors hold, $p_1, s_1 > 0$. Then the conclusions of Theorem 1.7 also hold.*

Remark. It is worth translating our last result into the language of probability theory (Lancaster, 1969, Chapter 9). Let there be n different possible states numbered as $i = 1, 2, \ldots, n$. Let $M = (m_{ij})$ be the matrix of *transition probabilities*, that is, where $m_{ij} \geq 0$ is the conditional probability that the system goes from state j to state i in one step. Then $\sum_i m_{ij} = 1$, $j = 1, \ldots, n$. Let x_t be a probability distribution vector where entry $x_{i,t} \geq 0$ is the probability that the system is in state i in time t. Obviously, $\sum_i x_{i,t} = 1$ and $x_{i,t} = \sum_j m_{ij} x_{j,t-1}$. Using matrix notations, $\mathbf{1}^{\mathbf{T}} M = \mathbf{1}^{\mathbf{T}}$, $x_t = M x_{t-1}$ and $\mathbf{1}^{\mathbf{T}} x_t = 1$. Theorem A.7b states that an acyclic (homogeneous) Markov chain is *ergodic*: $\lim_{t \to \infty} x_{j,t} = x_j^{\circ}$, $j = 1, \ldots, n$.

For a negative dominant root ($\lambda_1 = -1$), a very simple cycle-model is obtained: $s_1 = -M s_1$ implies $x_t \approx -x_{t-1}$, $x_t \approx x_{t-2}$, 2-cycle. There is "only" one problem: in contrast to the case $\lambda_1 = 1$, we have no reason to suppose that $\rho(M) = 1$ typically holds with $\lambda_1 = -1$.

Viability

We have already mentioned in the Introduction that stability is only an asymptotic qualification and we often need analyze *transient* processes. The notion of viability applies here. In the simplest case *viability* means that the state remains in the so-called *viability domain*. Symbolically, $x_t \in \mathcal{P}$ for every t where $\mathcal{P}$ is a suitable subset of $\mathbf{R}^n$, for example, $\mathcal{P} = \mathbf{R}_+^n$.

In Appendix A an important generalization of the length of a vector is introduced: *vector-norm*. With its help we can frequently give an estimation on those initial states which generate a viable trajectory. Let us assume that the fixed point x° is an interior point of $\mathcal{P}$, that is, there exists a positive number r, such that points x satisfying $||x - x^\circ|| < r$ are contained by $\mathcal{P}$. In formula: ball $B(x^\circ, r) \subseteq \mathcal{P}$.

Theorem 1.8. *(Martos, 1990.) a) If there exists a fixed point and if for an appropriate vector-norm and the induced matrix-norm,*

$$(1.25) \qquad ||M|| \leq 1 \quad (\text{respectively,} \quad ||M|| < 1),$$

then iteration (1.10) is Lyapunov (respectively, asymptotically) stable.

b) Under the assumption of point a) let us assume that for an appropriate $r > 0$, $B(x^\circ, r)$ is contained by the viability domain $\mathcal{P}$. Then every $x_0 \in B(x^\circ, r)$ generates a viable path.

Proof. The norm is helpful in proving stability (see Section 3.1 below).

a) We return to the notations of the deviation system: By (1.15^d), $x_t^\mathrm{d} = M x_{t-1}^\mathrm{d}$; by (A.17) and (1.25),

$$(1.26) \qquad ||x_t^\mathrm{d}|| = ||M x_{t-1}^\mathrm{d}|| \leq ||M|| \, ||x_{t-1}^\mathrm{d}|| \leq ||x_{t-1}^\mathrm{d}||.$$

Thus for every $\varepsilon > 0$, it is sufficient to choose $\delta = \varepsilon$, and $||x_0^\mathrm{d}|| < \delta$ implies $||x_t^\mathrm{d}|| < \varepsilon$. A similar proof yields asymptotic stability.

b) By $||x_t^\mathrm{d}|| \leq ||x_0^\mathrm{d}||$, $x_0 \in B(x^\circ, r)$ implies $x_t \in B(x^\circ, r)$. ∎

The next problem is useful as a simplified summary.

Problem 1.10. Scalar equation. Review all the theorems for the linear constant-coefficient scalar case $x_t = \lambda x_{t-1}$.

In case of $n = 1$, Figure 1.2 displays the sign-alternating (to be called oscillatory) paths with $M = -1.1$, -1 and -0.9; Figure 1.3 displays the constant-sign (to be called oscillation-free) paths with $M = 1.1$, 1 and 0.9. It can be seen that unstable, cyclic and stable paths appear.

1.3 PLANAR LINEAR SYSTEMS

We have seen in the solution to Problem 1.10 how simple the theory of time-invariant first-order scalar difference equation is. As a next step, we shall now consider constant-coefficient linear systems in two dimensions (*planar systems*). We have already met them, discussing complex roots. Planar systems will serve us well in many occasions, as the second simplest case after the scalar ones.

Basic concepts

Let M be a 2×2 matrix:

$$M = \begin{pmatrix} m_{11} & m_{12} \\ m_{21} & m_{22} \end{pmatrix}.$$

We shall need the quadratic characteristic polynomial $P(\lambda)$ of the planar system:

$$(1.27) \qquad P(\lambda) = \lambda^2 - \omega\lambda + \vartheta$$

where

$$(1.28) \quad \omega = \operatorname{tr} M = m_{11} + m_{22}, \quad \vartheta = \det M = m_{11}m_{22} - m_{12}m_{21}.$$

We shall speak of (asymptotic) *oscillation* if there is no date after which all the deviation variables keep their signs for ever.

We shall distinguish two cases:

a) There is a *degenerate oscillation* if after a transitional period both variables change signs in each time-period.

b) There is a *regular oscillation* if the two variables do not change signs in each time-period, but oscillate.

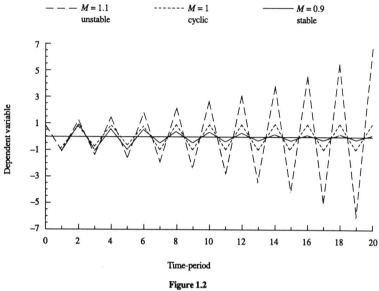

Figure 1.2
Oscillations

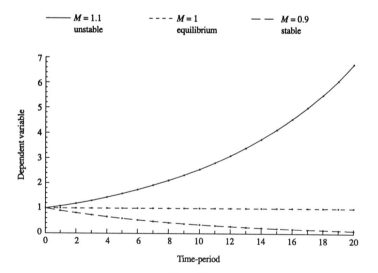

Figure 1.3
Oscillation-free paths

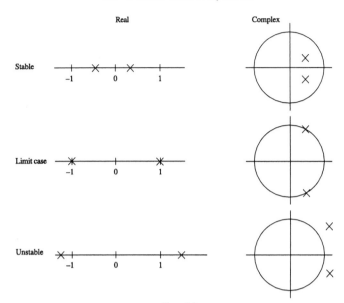

Figure 1.4
Classification of characteristic roots

Classification

First we shall classify the planar linear systems by their characteristic roots. We shall see in nonlinear systems (in Chapters 3 and 4) that different initial states of the same system can generate different types of paths (oscillation-free versus oscillating or stable versus unstable). In our linear system with two variables, however, the classification is independent of the initial states, apart from an insignificant set of them.

Theorem 1.9. *Apart from atypical cases, constant-coefficient planar linear systems can be classified as follows. We shall assume* $|\lambda_2| \le |\lambda_1|$.

a) The dominant root is positive, $|\lambda_2| < \lambda_1$: *asymptotically oscillation-free.*

b) The dominant root is negative, $|\lambda_2| < -\lambda_1$: *degenerate oscillation.*

c) The dominant roots are complex, $|\Re\lambda_1| < |\lambda_1|$: *regular oscillation.*

Remark.   We have no space to discuss each possible case of stability and (no)-oscillation. We only note that this multitude of cases which characterizes difference equations is the reason that mathematicians generally prefer differential equations to difference equations (see Chapter 5 below).

Proof.   Assume the existence of a characteristic basis. (1.21) has now two terms: $x_t = \xi_1 \lambda_1^t s_1 + \xi_2 \lambda_2^t s_2$, and because of the dominance of λ_1, $x_t \approx \xi_1 \lambda_1^t s_1$. In case of real roots, at large t's the sign is independent of t for a) and alternates for b). c) By (1.22).   ∎

Figure 1.4 presents a slightly different classification.

Problem 1.11.   a)   Enumerate the missing atypical cases. b) What is the qualitative difference between systems with the following two pairs of characteristic roots: (1;.5) and (1;-.5)?

Oscillations

We continue the analysis of oscillations.

Theorem 1.10.   *Using the notations (1.27)–(1.28), degenerate oscillation occurs if and only if*

$$(1.29) \qquad\qquad \omega^2 \geq 4\vartheta$$

and

$$\omega \leq 0.$$

Proof.   According to the relation between the coefficients and the roots of any quadratic equation,

$$(1.30) \qquad\qquad \lambda_1 + \lambda_2 = \omega \quad \text{and} \quad \lambda_1 \lambda_2 = \vartheta.$$

(1.29) is equivalent to the fact that both roots of $P(\lambda)$ are real. $\lambda_1 + \lambda_2 = \omega$ and $\omega \leq 0$ ensure that there is a negative dominant root. ∎

Example 1.11.   Degenerate versus regular oscillation. a) A matrix with a negative dominant diagonal $-m_{11} > |m_{12}|$, $-m_{22} > |m_{21}|$ implies degenerate oscillation. b) The 4-phase pendulum regularly oscillates (with a 4-cycle in Problem 1.1).

We shall analyze regular oscillation in more detail. Then by interpolation, the solution can be made continuous in time, since functions $\cos \varphi t$ and $\sin \varphi t$ in (1.22) are defined for any real t. One can define also a period of cycle for any positive real number. Thus it is demonstrated that for an oscillation interpolated in continuous time, the direction of the deviation of state always returns to the initial one, and the time elapsing between the two events is the period with continuous time.

We shall call a planar system *cyclic in continuous time* if it oscillates and not only the direction but the state as well returns to its initial position. Within this case two subcases will be distinguished in discrete time: α) *cycle* and β) *quasi-cycle*. Cycles have already been defined in Section 1.1.

At the beginning of the 20th century, Weyl studied the simplest case of the quasi-cycle, namely, the rotation with an 'irrational angle' φ. Let ψ_t be a real number, representing the state of the system at date t on the unit circle line by the angle and let φ/π be an irrational number. Then $\psi_t = \psi_{t-1} + \varphi$, $\psi_0 = 0$. He discovered that this map generates a uniform distribution, that is, in the long run, the number of points falling in any given interval of the circle line is proportional to the length of that interval (Pólya and Szegő, 1924, Volume II, Part II, Section 4.1).

For better understanding, we present a concrete example from astronomy. (i) In a first approximation the Earth's orbit around the Sun has a so-called composite cycle of period $4 \cdot 365 + 1$ days, which led Julius Caesar to introduce every fourth year as a leap-year from 44 B.C. (ii) In a second approximation the Earth's orbit around the Sun is quasi-cyclic, because the ratio of the length of the astronomical year to the length of day is not a rational number (with a small denominator). This led to a sophisticated system of leap-years introduced by Pope Gregory XIII in 1582: years which can be divided by 100 are not leap-years unless they can also be divided by 400.

The following theorem and example contain the oscillation and stability conditions in terms of the coefficients rather than the roots, respectively.

Theorem 1.11. a) *A planar linear system regularly oscillates if and only if the negation of (1.29) holds:*

$$\omega^2 < 4\vartheta.$$

b) For a regular oscillation, period P and the dampening factor Φ of the continuous solution are independent of the initial state and given by

(1.31) $P = \dfrac{2\pi}{\varphi}$ where $\varphi = \arccos\left(\dfrac{\omega}{2\sqrt{\vartheta}}\right)$

and

(1.32) $\Phi = \dfrac{1}{\sqrt{\vartheta}}.$

c) A regularly oscillating system is cyclic in continuous time if and only if in addition to negation of (1.29), the following condition holds:

(1.33) $\vartheta = 1.$

Remark.   Similarly to the foregoing theorem we can make related statements concerning unexamined cases. For example, a regular oscillation is stable if and only if $\vartheta < 1$.

Proof.   a) Complex roots appear with negative discriminants.

b) Let the two roots be $\lambda_{1,2} = (\cos\varphi \pm i\sin\varphi)/\Phi$ where φ and Φ are positive real numbers and i is the imaginary unit root. Substitution of the pair of roots into (1.30) yields $2\cos\varphi = \omega\Phi$ and $1/\Phi = \sqrt{\vartheta}$. This proves (1.31)–(1.32).

c) Obviously, a regular system is cyclic in continuous time if the dampening factor is equal to unity, that is, by (1.32), if (1.33) holds.∎

Example 1.12.   Stability conditions (Samuelson, 1947). A planar linear dynamical system is stable if and only if the coefficients of the characteristic polynomial (1.27)–(1.28) satisfy the following three conditions: $1 + \omega + \vartheta > 0$, $1 - \omega + \vartheta > 0$ and $\vartheta < 1$.

a) In case of complex roots: $\omega^2 < 4\vartheta$, (1.32): $0 < \vartheta < 1$. $1 \pm \omega + \vartheta > 1 - 2\sqrt{\vartheta} + \vartheta = (1 - \sqrt{\vartheta})^2 > 0$.

b) In case of real roots, we shall again rely on the relations among roots and coefficients [(1.30)]. Discussion of several cases reveals that $-1 < \lambda_1 < \lambda_2 < 1$ is equivalent to $P(1) > 0$, $P(-1) > 0$ and $\vartheta < 1$.

There are two basic difficulties with linear cycle models: a) regular quasi-cyclic and cyclic systems arise only for exceptional parameter values; see ϑ in (1.33); b) the largest deviation (the so-called amplitude) is uniquely determined by the initial state. This twin-problem can only be solved in nonlinear models (see Chapters 3 and 4).

1.4* LINEAR CONTROL SYSTEMS

In this Section we shall briefly discuss the linear class of control systems defined at the end of Section 1.1 (see Aoki, 1976 and Martos, 1981).

Controllability

Let $x \in \mathbf{R}^n$ and $u \in \mathbf{R}^m$ be the *state vector* and the *control vector*, respectively. Let A and B be the $n \times n$ *system matrix* and the $n \times m$ *input matrix*, p and q be n- and m-vectors, respectively. We can now describe the time-invariant linear system's
State space equation in discrete time:

$$(1.34) \qquad x_t = Ax_{t-1} + Bu_t + p, \qquad t = 1, 2, ..,$$

the initial state x_0 is given. We may assume that the dimension of the control vector is at most as large as that of the state vector: $m \leq n$ (why?).

The next example presents a control system where $m < n$.

Example 1.13. $n = 2 > m = 1$. $y_t = \alpha_1 y_{t-1} + \alpha_2 y_{t-2} + \beta u_t$ where y_t and u_t are scalar variables. Let $x_t = (y_t, y_{t-1})$, then the transformation presented in Example 1.2 yields

$$x_{1,t} = \alpha_1 x_{1,t-1} + \alpha_2 x_{2,t-1} + \beta u_t \quad \text{and} \quad x_{2,t} = x_{1,t-1}.$$

Hence

$$A = \begin{pmatrix} \alpha_1 & \alpha_2 \\ 1 & 0 \end{pmatrix}, \qquad B = \begin{pmatrix} \beta \\ 0 \end{pmatrix}.$$

We shall prove

Theorem 1.12. *(Kalman, 1960.) A linear control system (A, B) is controllable if and only if the following rank condition holds:*

$$(1.35) \qquad r[B, AB, \ldots, A^{n-1}B] = n$$

where the matrices $B, AB, \ldots, A^{n-1}B$ of order $n \times m$ form an $n \times mn$ hypermatrix. Then the objective can be reached within n time-periods.

Before spelling out the proof, we present two extreme examples.

Example 1.14. Controllability of a simple control system. For $A = I$, (1.35) reduces to $r(B) = n$, that is, $m \geq n$. Choosing n independent columns from B, we obtain the same number of state variables as that of control variables, and the aim can be achieved without delay. Assume B is an $n \times n$ matrix. Then $u_1 = B^{-1}(x_1 - x_0 + p)$.

Example 1.15.* Controllability of a scalar control system. For $m = 1$, (1.35) simplifies into the cyclic condition, well-known from the geometrical view of Jordan normal form of matrices (discussed in the subsection on multiple characteristic roots above): the vectors $b, Ab, \ldots, A^{n-1}b$ should be independent, moreover, they should form a basis.

Proof. *Necessity.* Assume that the system is controllable and $p = 0$. Then there exists a series $u_1, \ldots, u_T$ for which $x_0 = x^1$ and $x_T = x^2$ hold. Using a well-known method of undetermined multipliers, multiply both sides of (1.34) by A^{T-t}:

$$A^{T-t}x_t = A^{T-t+1}x_{t-1} + A^{T-t}Bu_t, \qquad t = 1, \ldots, T.$$

Observe that the expression of the L.H.S. of equation t reappears at the R.H.S. of equation $t+1$. Thus adding up the T equations, they cancel out and we obtain

$$(1.36) \qquad\qquad x_T = A^T x_0 + \sum_{t=1}^{T} A^{T-t}Bu_t.$$

Let $x_0 = 0$. According to (1.36), any vector x_T can be represented as the corresponding sum, that is, the rank condition holds. By the Cayley–Hamilton theorem (Theorem A.4 in the book), there exists an integer T, $T \leq n$, such that (1.36) holds.

Sufficiency. If the rank condition holds, then for an appropriate series $\{u_t\}$, (1.36) holds, that is, the system is controllable. ∎

Remark. It is worth mentioning that the two terms at the R.H.S. of (1.36) have a transparent meaning: a) the first term is the homogeneous solution for x_0 (at $u = 0$), b) the second term is a particular solution belonging to $x_0 = 0$.

The following well-known example illustrates this technique.

Example 1.16. The formula for the sum of the geometric progression: $q \neq 1$, $s_t = s_{t-1} + q^t$, $s_0 = 1$. With successive substitution, $s_t = 1 + q + \ldots + q^t$. Multiplying both sides by q: $qs_t = q + q^2 + \ldots + q^{t+1}$. Deducting qs_t from s_t, $s_t = (q^{t+1} - 1)/(q - 1)$.

Observability

It occurs quite often that one cannot observe the state vector in full, still, one wants to control system (1.34). This case is formalized below. Let $y \in \mathbf{R}^z$ be the *observation vector* and C be a $z \times n$ matrix. We have the
Observation equation

$$(1.37) \qquad\qquad y_t = C x_t.$$

We shall call system (A, C) *observable* if any initial state x_0 can be determined from a finite sequence of observations $y_0, y_1, \ldots, y_T$.

It can be shown that the problem of observability is a dual of that of controllability. Thus Theorem 1.12 implies

Theorem 1.13. *(Kalman, 1960.)* The observation system (A, C) *is observable if and only if the following rank condition holds:*

$$(1.38) \qquad\qquad r[C, CA, \ldots, CA^{n-1}] = n.$$

The duals of Examples 1.14 and 1.15 are as follows.

Example 1.17. Observability of a simple control system. For $A = I$, (1.38) reduces to $r(C) = n$, that is, $z \geq n$. Choosing n independent rows from C, we obtain the same number of state variables as information variables, and the reconstruction can be achieved without delay.

Example 1.18.* Observability of a scalar control system. For $m = 1$, (1.38) simplifies into the cyclic condition, well-known from the Jordan normal form of the matrices: vectors $c, cA, \ldots, cA^{n-1}$ should be independent, moreover, they should form a basis.

From now on we shall always assume that system (A, B, C) is controllable and observable.

Feedback and stabilizability

We shall speak of *linear feedback* if the control is a linear function of the state:

$$(1.39) \qquad\qquad u_t = -K x_{t-1} + q.$$

The entries k_{ij} of matrix K are called *reaction coefficients* (or adjustment speeds), since they show the strength of control of i with respect to state variable of j. System (1.34) is called *stabilizable by the linear feedback* (1.39), for short, *stabilizable* if there exists an $m \times n$ matrix K, such that the resulting matrix $M = A - BK$ is stable: $\rho(M) < 1$. We state without proof

Theorem 1.14. *(Kalman, 1960.) If the control system (A, B) is controllable, then it is stabilizable.*

As an illustration, we present

Example 1.19. Continuation of Example 1.14. For the simple dynamical system $x_t = x_{t-1} + Bu_t$, controllability (B^{-1} exists) yields stabilization: $K = B^{-1}$.

Decentralized stabilization

The concepts of *decentralized stabilization and stabilizability* play a central role in numerical analysis as well as in economic control theory. If the system is large, then the centralization of all information into a single center may be impractical if not impossible. For our applications (see Sections 2.2, 2.3 and 6.3), it is sufficient to consider the simple case, when $m = n$, $A = I$, hence B is invertible (Example 1.14) and the feedback matrix K is diagonal: $K = \langle k \rangle$, or can be made diagonal with appropriate changes of rows. Totally decentralized feedback can be visualized as follows: there are n controllers. Controller i observes his own state variable $x_{i,t-1}$ and reacts via k_i on $u_{i,t}$. Now (1.34) and (1.39) simplify into

$$(1.40) \qquad x_t = x_{t-1} + Bu_t + p,$$
$$(1.41) \qquad u_t = -\langle k \rangle x_{t-1} + q;$$

respectively, that is,

$$(1.42) \qquad x_t = (I - B\langle k \rangle)x_{t-1} + p + Bq.$$

Hence $M = I - B\langle k \rangle$ and $w = p + Bq$.

Before presenting our results on decentralized stabilization, we introduce a definition which is slightly stronger than the invertibility (regularity) of matrix B. A matrix B is called *strongly regular* if, with the simultaneous changes of rows and columns, it can be transformed in such a way that the resulting blocks $B_r = (b_{ij})_{1 \le i,j \le r}$ are also invertible, $r = 1, 2, \ldots, n$.

Applying a theorem by Fuller and Fisher (1958) yields

Theorem 1.15. (*McFadden, 1969.*) *Every simple control system (1.40) with strongly regular input matrix B is decentralizedly stabilizable.*

Sketch of the proof. The proof is similar to that of ε-triangularization of a matrix (see Problem 1.7 above and part a) of Proof of Theorem 5.9 below). Controller i essentially only affects the changes in states of controllers j, $(j = 1, 2, \ldots, i-1)$, and the order of magnitude of the reaction coefficient k_i is ε^i where ε is sufficiently close to 0. Unfortunately, the resulting convergence is very slow. ∎

Remark. At this point we make a statement with implications for numerical analysis as well as for economics. Using the language of numerical analysis, an iterative solution of (1.40)–(1.41) with fast convergence is algorithmically superior to the direct solution of (1.13) via inverting $I - M$. Bodewig (1959) and Varga (1962) give a very vivid historical description, going back to a paper by Gauss in 1823.

The next problem is the simplest illustration of the distinction mentioned in Theorem 1.15.

Problem 1.12. Centralized versus decentralized stabilizability. Prove that for $\alpha, \beta < 0$, matrix

$$B = \begin{pmatrix} 0 & \alpha \\ \beta & 0 \end{pmatrix}$$

is a) invertible, b) not strongly regular, c) can be stabilized in a centralized way but d) cannot be stabilized in a decentralized way.

For further analysis, we shall need the coordinate-wise form of (1.40):

$$x_{i,t} = x_{i,t-1} + b_{ii}u_{i,t} + \sum_{j \neq i} b_{ij}u_{j,t} + p_i, \quad i = 1, \ldots, n.$$

We shall now consider a special class of matrices (called *M-matrices*, Young, 1971, Section 2.7) which plays an important role in numerical analysis as well as in economic applications (Young, 1971, Section 2.6 and Simonovits, 1981b). The diagonal entries of the matrix are positive (unitary) and the off-diagonal entries are nonpositive:

$$(1.43) \qquad b_{ii} = 1 \quad \text{and} \quad b_{ij} \leq 0, \quad i \neq j.$$

Let us introduce the *matrix of cross-effects*:

$$(1.44) \qquad\qquad N = I - B \geq 0,$$

and assume that N is irreducible, moreover, *the own-effects dominate the cross-effects*:

$$-\sum_{i \neq j} b_{ij} < 1, \quad j = 1, \ldots, n.$$

In vector form:

$$(1.45) \qquad\qquad \mathbf{1^T} N < \mathbf{1^T} \quad \text{where} \quad \mathbf{1^T} = (1, \ldots, 1).$$

As is mentioned in Corollary b) to Theorem A.9, (1.45) is equivalent to the stability of N: $\rho(N) < 1$.

We turn to the dynamics. First of all, introduce notation

$$(1.46) \qquad\qquad M = I - B\langle k \rangle = I - \langle k \rangle + N\langle k \rangle.$$

Notice that M has advantageous properties if the reaction coefficients are *damped*:

$$(1.47) \qquad\qquad 0 < k \leq \mathbf{1}.$$

Lemma 1.1. *Assume that the positive own-effects dominate the negative cross-effects [(1.43) and (1.45)], and the feedback is damped [(1.47)]. Then M is nonnegative, irreducible and its spectral radius is a decreasing function of any reaction coefficient, thus it is less than unity: matrix M is stable.*

Proof. By (1.47), M in (1.46) is nonnegative and irreducible. Let the positive dominant root of M be ρ, and the corresponding positive characteristic vector be s (Theorem A.7). Substitute (1.46) into $\rho s = Ms$:

$$(1.48) \qquad\qquad \rho s = (I - \langle k \rangle + N\langle k \rangle)s,$$

and rearrange it in such a way that a matrix of positive entries is obtained, being an increasing function of $\langle k \rangle$. With appropriate rearrangement, $(1 - \rho)^{-1} s = \langle k \rangle^{-1}(I - N)^{-1}s$ where $(I - N)^{-1} > 0$ (see Theorem A.7e) appears on the R.H.S. Applying Theorem A.7d, the spectral radius of matrix $\langle k \rangle^{-1}(I - N)^{-1}$ is an increasing function of any of its entries. For a fixed i, if k_i increases (that is, k_i^{-1} decreases), then $(1 - \rho)^{-1}$ decreases, hence ρ also decreases. For $k = 0$, $\rho = 1$ holds, for $0 < k \leq 1$, $\rho < 1$ holds. (The case of semipositive ks, where some but not all k_i's are positive, is skipped.) ∎

Due to the simple structure of the model, the results of Section 1.2 can be directly applied to stability without oscillation. Corollary to Theorem 1.6 implies

Theorem 1.16. *Retain the assumptions of Lemma 1.1.*

a) The decentralized feedback is stable, its dampening factor is given by

$$(1.49) \qquad \Phi_k = \frac{1}{\rho(I - \langle k \rangle + N\langle k \rangle)}.$$

b) Its maximal dampening factor is achieved for $k = 1$, and its value is given by

$$(1.50) \qquad \Phi_{\mathbf{1}} = \frac{1}{\rho(N)}.$$

c) The feedback is typically (asymptotically) oscillation-free.

Remark. Putting aside condition $0 < k \leq 1$, the convergence can generally be improved. This phenomenon is well-known in numerical analysis of iterative solution of systems of linear algebraic equation and is called *overrelaxation* (Young, 1971, Chapter 4).

We can sharpen our results by introducing the class of *P-cyclic matrices* (see Appendix A) where matrix N_Q is an $n_Q \times n_{Q+1}$ matrix, $\sum_Q n_Q = n$.

$$(1.51) \qquad N = \begin{pmatrix} 0 & N_1 & 0 & \ldots & 0 \\ 0 & 0 & N_2 & \ldots & 0 \\ \vdots & \vdots & \vdots & \ddots & \vdots \\ N_P & 0 & 0 & \ldots & 0 \end{pmatrix}.$$

We shall call the linear feedback (1.41) *uniform* if all reaction coefficients are the same:

$$(1.52) \qquad k = \kappa \mathbf{1}.$$

For cyclic matrices of cross-effects and uniform feedback, Theorem 1.16 can be sharpened.

Theorem 1.17. a) *Assume that the positive own-effects dominate the negative cross-effects [(1.43) and (1.45)], the matrix of cross-effects is P-cyclic [(1.51)] and the feedback is uniform [(1.52)]. Then the dampening factor is a decreasing function of the uniform reaction coefficient $\kappa > 1$.*

b) *The maximal dampening factor is achieved for $\kappa = 1$.*

c) *Assume that matrix N is 2-cyclic and the decentralized feedback is unconstrained. Then the globally maximal dampening factor is also achieved at $k = 1$.*

Proof.   a) By (1.46) and (1.52), $M = (1 - \kappa)I + \kappa N$. Let $\vartheta = \rho(N)$. By Theorem A.8a, the loci starting from the P dominant roots are given by $\lambda_Q = 1 - \kappa + \kappa \vartheta \varepsilon^{Q-1}$, $Q = 1, 2, \ldots, P$. It is easy to see that for $\kappa > 1$, $\lambda_1 > 0$ ceases to be dominant, and is replaced by a characteristic root with index $q = [(P + 1)/2]$ (where $[\cdot]$ is the integer part of a real number). Indeed, for even P, $\lambda_q = 1 - \kappa - \kappa \vartheta < -\vartheta$. For odd P, we have a right-angle triangle with perpendicular side lengths $\kappa - 1$ and $\kappa \vartheta \varepsilon^q$, hypotenuse with length $|\lambda_q|$. In both cases $|\lambda_q| > \vartheta$.

b) It is an obvious corollary of Theorem 1.16 and part a).

c) Having renounced both dampedness and uniformity, the proof becomes so complicated that we omit it. The basic idea rests on the following transformation:

$$(1.53) \qquad x = [(\lambda - 1)I + \langle k \rangle]^{-1} \langle k \rangle N x.$$

For $P = 2$ and $k = 1$, in addition to a positive characteristic root ρ, the negative $-\rho$ is also a characteristic root. Up to now it is an open question if statement c) is true for $P > 2$.   ∎

Example 1.19. Illustration. For $n = 2$ and 2-cyclic system $(n_{11} = n_{22} = 0)$ and uniform feedback, both characteristic roots are real, for $k = 0$, both are equal to 1, for $k = 1$, $\pm\sqrt{b_{11} b_{22}}$ is the optimum.

2. DISCRETE-TIME LINEAR MODELS

In this Chapter we shall analyze *discrete-time linear* dynamic economic models. We shall explain the meaning of our two adjectives. (i) In *discrete-time* models the unit of time is one year or one quarter or one month, but at extreme it may be of 30 years (Appendix B). Due to the inherent lags in these models, now the discrete-time approach is preferable to the continuous-time approach. (ii) Roughly speaking, in *linear* models output is proportional to input. We shall consider nonlinear or continuous-time models (Chapters 4 and 6, respectively) later on. In Section 2.1 the *accelerator–multiplier* model of the trade cycle is discussed. In Section 2.2 we shall investigate the *control* of a multisector economy by *stock signals*. Section 2.3* discusses the role of *expectations in a linear control model*. The first model is a macro-model, while the second and third ones are not. It is noteworthy that in all the three models prices play a subordinate role (as is the case in Disequilibrium Theory and Non-Price Control) and each needs a nonlinear extension (Chapter 4).

2.1 LINEAR MODEL OF ACCELERATOR–MULTIPLIER

First we shall present the *simple* model of Hicks (1950), then we shall outline the corresponding *composite* model. As is usual in macroeconomics, we shall work with volumes (indices calculated at constant prices). Let Y_t, C_t and I_t be the volumes of output (GDP), consumption and net investment in time-period t, respectively. (More precisely, I_t is accumulation which includes investment *and* inventory accumulation.) The economy is closed.

SIMPLE MODEL

Volumes

In a closed economy there is an identity among the three variables: output is equal to investment and consumption.
GDP identity

$$(2.1) \qquad\qquad Y_t = I_t + C_t.$$

In 1917 J. M. Clark introduced the investment *accelerator*, which is the proportionality factor (β) between the change in output and the investment. Hicks (1950) added the autonomous investment I_t^A.
Investment function

$$(2.2) \qquad\qquad I_t = I_t^A + \beta(Y_{t-1} - Y_{t-2}).$$

Since Keynes (1936), consumption has frequently been described as the linear function of previous income (that is, output).
Consumption function

$$(2.3) \qquad\qquad C_t = C_t^A + \gamma Y_{t-1}$$

where $0 < \gamma < 1$ is the *marginal rate (propensity) of consumption*, yielding the famous *multiplier* $1/(1 - \gamma)$.

For given paths I^A, C^A, given parameters β, γ, and given initial values Y_{-2}, Y_{-1}, the paths I, C and Y are uniquely determined.

If I^A and C^A are regular, that is, they expand by the same time-invariant *growth factor* $\Gamma > 1$, that is, if

$$(2.4) \qquad\qquad I_t^A = i^A \Gamma^t \quad \text{and} \quad C_t^A = c^A \Gamma^t,$$

then the corresponding I, C and Y are also regular with the same growth factor Γ.

Relative values

Under condition (2.4) it is easier to study the problem in a relative system where the original variables and several parameters are divided by the growth trend.

Relative variables

$$(2.5) \qquad y_t = \frac{Y_t}{\Gamma^t}, \quad i_t = \frac{I_t}{\Gamma^t} \quad \text{and} \quad c_t = \frac{C_t}{\Gamma^t}.$$

Relative parameters

$$(2.6) \qquad i^A = \frac{I_t^A}{\Gamma^t} \quad \text{and} \quad c^A = \frac{C_t^A}{\Gamma^t}.$$

Having introduced notation $\psi = 1/\Gamma$, we have the
Relative equations

$$(2.1') \qquad y_t = i_t + c_t,$$
$$(2.2') \qquad i_t = i^A + \beta\psi(y_{t-1} - \psi y_{t-2}),$$
$$(2.3') \qquad c_t = c^A + \gamma\psi y_{t-1}.$$

Basic equation

We have three equations and three variables. It is useful to eliminate the superfluous variables and equations. To transform our system $(2.1')$–$(2.3')$ into a canonical form, we have two options: either to have a planar system of first-order equations or to have a scalar second-order equation. We choose the second variant. Substitute $(2.2')$ and $(2.3')$ into $(2.1')$. After some manipulation we obtain the
Basic equation

$$(2.7) \quad y_t = i^A + c^A + (\beta + \gamma)\psi y_{t-1} - \beta\psi^2 y_{t-2}, \qquad t = 0, 1, 2, \dots$$

where y_{-2} and y_{-1} are given initial values.

Having the dynamics of the *basic variable* y_t, the dynamics of the remaining variables (i_t and c_t) are easily determined from $(2.2')$ and $(2.3')$.

We will state two theorems: one on the fixed point, the second on stability and oscillation.

Theorem 2.1. *(Hicks, 1950.) There exists a unique equilibrium in the simple Hicksian economy:*

$$(2.8) \quad y^\circ = \frac{i^A + c^A}{1 - \psi(\beta + \gamma) + \psi^2\beta} \quad \text{where} \quad 1 - \psi(\beta + \gamma) + \psi^2\beta > 0.$$

The rather complex classification in Theorems 1.9–1.11 now becomes quite simple:

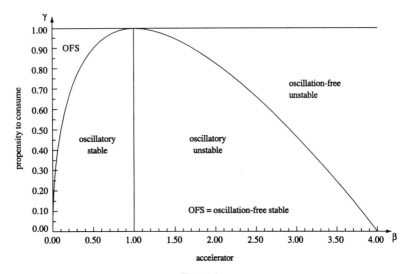

Figure 2.1
Hicksian classification

Theorem 2.2. *(Hicks, 1950.) Suppose the simple Hicksian system is stagnating:* $\psi = 1$.

a) The system oscillates if and only if

$$(2.9) \qquad\qquad \gamma < 2\sqrt{\beta} - \beta.$$

b) The system is stable if and only if

$$(2.10) \qquad\qquad \beta < 1.$$

Figure 2.1 illustrates Theorem 2.2.

Remark. In realistic conditions $\beta \approx 2$, $\gamma \approx 0.75$, therefore oscillation and stability are inconsistent. We have to replace the linear deterministic model with either a nonlinear deterministic (Section 4.2) or a linear stochastic model (Chapter 7).

Proof. Substitute the basic solution $y_t = y_0 \lambda^t$ into the homogeneous part of (2.7). This yields the following quadratic characteristic polynomial: $P(\lambda) = \lambda^2 - (\beta + \gamma)\lambda + \beta$.

We shall separate the cases of a) complex and b) real characteristic roots. Ad a) The negativity of the discriminant provides

(2.9). Then the Remark to Theorem 1.11 yields the stability condition (2.10).

Ad b) We shall rely on the well-known relations among roots and coefficients (1.30). Since $\lambda_1 \lambda_2 = \beta > 0$ and $\lambda_1 + \lambda_2 = \beta + \gamma > 0$, λ_1 and λ_2 are positive. By Theorem 1.9b, there is no degenerate oscillation. ∎

Problem 2.1. Four possibilities. Choose parameter values such that each of the four possibilities, illustrated in Figure 2.1, occur while $\psi = 1$. Write a computer program on the output paths and run it for all the four cases ($i^A = 0$, $c^A = 1 - \gamma$, Blatt, 1983, p. 192). Figures 2.2–2.5 display paths for $\gamma = 0.75$, $\beta = 0.25$; 0.9; 1.5, 2.2, respectively with common initial conditions: $Y_{-2} = Y_{-1} = 1.2$.

Example 2.1. Hicksian control model. On the basis of Example 1.13, the Hicksian model can be described as a control model where the autonomous investment is the control variable.

COMPOSITE MODEL

Already Hicks (1950) studied the composite model where the investment or the consumption function is based on distributed lags:
Investment function with distributed lags

$$(2.11) \qquad i_t = i^A + \sum_{k>0} \psi^k \beta_k (y_{t-k} - \psi y_{t-k-1}).$$

Consumption function with distributed lags

$$(2.12) \qquad c_t = c^A + \sum_{k>0} \psi^k \gamma_k y_{t-k}, \quad \gamma_k \geq 0; \quad \sum_{k>0} \psi^k \gamma_k < 1.$$

Many more cases are now possible, however, this problem is outside the scope of the book. We shall return to the nonlinear version of the composite Hicksian model in Section 4.2.

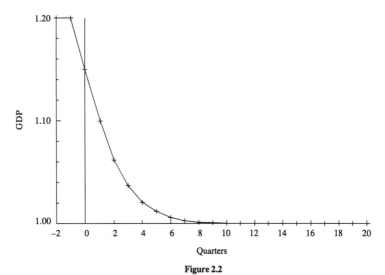

Figure 2.2
Stable Hicks path without oscillation

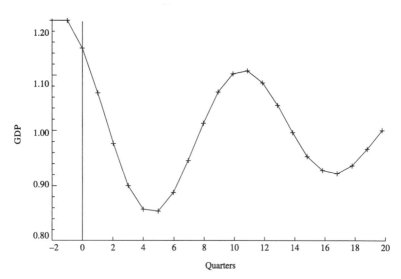

Figure 2.3
Stable Hicks path with oscillation

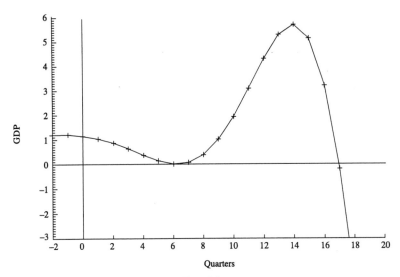

Figure 2.4
Unstable Hicks path with oscillation

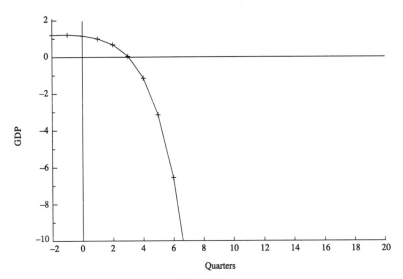

Figure 2.5
Unstable Hicks path without oscillation

2.2 LINEAR MODELS OF CONTROL
BY STOCK SIGNALS

In this Section we shall present two linear models of control by stock signals. The basic idea is borrowed from Kornai and Martos (1973), a good deal of papers in Kornai and Martos eds. (1981b) discuss this type of models. The simplest description is due to Martos (1990).

CONTROL BY OUTPUT STOCK SIGNALS

The model

We set out from an n-sector model where the interdependence of the different sectors is described by an open Leontief model (Leontief, 1941; Morishima, 1964 and Bródy, 1970), augmented by stocks (inventories). In contrast to the traditional inventory models, in this model (as well as in other members of the model family) we want to analyze the following issue: how can a whole economy function on the basis of decentralized control by stock signals? Simply speaking, we assume that each producer (sector) increases or decreases its output with respect to a given norm whether its own output stock is lower or higher than normal. Modelling a *just in time* (JIT) system, there are no input stocks in this model.

For the sake of simplicity, we assume an economy without long-run change. Let a_{ij} be the unit input requirement from sector i to sector j. Let $z_{i,t}$, $y_{i,t}$ and c_i be the *end-period output stock, output* of sector i in time-period t and the constant *consumption* from sector i, respectively. The corresponding matrix and vector notations are as follows: A, z_t, y_t and c. As is usual, it is assumed that A is a nonnegative, irreducible matrix with a spectral radius less than unity: $\rho(A) < 1$. With a suitable choice of units of measurements, this assumption is equivalent to $\mathbf{1}^T A < \mathbf{1}^T$ (Corollary to Theorem A.9). The dynamics of the model rests on a simple identity: change in stocks is equal to output less producers' inputs less consumption. The producers' inputs are proportional to outputs, the vector of final consumption is given. The model is closed by a simple linear decentralized control rule: the output of sector i is equal to the capacity y_i^* less reaction coefficient times own output stock.

The equations of the model

We can now present the equations of this simple model.
Change in output stocks

$$(2.13) \qquad z_t = z_{t-1} + (I - A)y_t - c.$$

Decentralized control of output

$$(2.14) \qquad y_t = y^* - \langle d \rangle z_{t-1}$$

where $\langle d \rangle = \langle d_i \rangle > 0$ is a diagonal matrix of the reaction coefficients.
Simple substitution of (2.14) into (2.13) yields the
Basic equation

$$z_t = (I - \langle d \rangle + A\langle d \rangle)z_{t-1} + (I - A)y^* - c.$$

Normal state

Before studying the dynamics of the basic equation, we shall investigate the properties of the *normal state* (or steady state) directly.

First of all, we shall make the following assumption. The full capacity output is capable to satisfy the final consumption:

$$(2.15) \qquad y^* > (I - A)^{-1}c.$$

Theorem 2.3. *For large enough capacities [(2.15)], the model controlled by output stock signals has unique normal control and state vectors which are positive:*

$$(2.16) \quad y^\circ = (I - A)^{-1}c > 0 \quad \text{and} \quad z^\circ = \langle d \rangle^{-1}[y^* - (I - A)^{-1}c] > 0.$$

Proof. Substitute $z_t = z_{t-1} = z^\circ$ and $y_t = y^\circ$ into (2.13)–(2.14), whence a simple calculation yields the equalities in (2.16). By $\rho(A) < 1$, Theorem A.7e and $c > 0$, $y^\circ > 0$ holds; by (2.15), $z^\circ > 0$. ∎

Stability without oscillation

Making the correspondences $x \to z$, $u \to y$, $M \to A$, $w \to c$ and $k \to d$, Theorem 1.16 applies. We repeat the previous definition, namely,
Feedback is damped:

$$(2.17) \qquad 0 < d \leq 1.$$

Theorem 2.4. *Assume damped feedback: (2.17).*

a) The control by output stock signals is stable.

b) Its maximal dampening factor is achieved for $d = \mathbf{1}$, and its value is given by

$$\Phi_{\mathbf{1}} = \frac{1}{\rho(A)}.$$

c) The feedback is typically (asymptotically) oscillation-free.

Remarks.   1. If assumption (2.17) is put aside, then the dampening factor can be generally further increased.

2. Atkinson's (1969) warning, that the economists generally neglect the quantitative aspects of dampening coefficient and are satisfied with the qualitative aspect of stability, is still timely. Our model is promising, yielding quite fast convergence.

3. Observe that a stable control by stock signals steers the system to the stationary stock vector. Without this iteration, the value of z° can only be determined in a centralized control.

4. Originally Kornai and Martos (1973) studied a model in continuous time and with permanent growth. In addition to stock signals, their producers also observed the changes in inputs and consumption, undermining the decentralized character of the control (Martos, 1990). Kornai and Simonovits (1977) and (1981) considered a discrete-time model and the *on-line* (that is, real time) decentralization of control was achieved by *off-line* (that is, imaginary time) centralization of the calculation of the norms: the latter was done in (2.16), the former in

$$(2.14') \qquad\qquad y_t - y^{\circ} = -\langle d \rangle (z_{t-1} - z^{\circ}).$$

Martos (1990) returned to the stagnating economy and anticipating the idea of the present model, set up a model with totally (off-line and on-line) decentralized control by stock signals.

5. It is noteworthy that Lovell (1962) and Bródy (1981) constructed similar models which were based on sales expectations (Section 2.3*) and price control (Section 6.3 below), respectively.

6. Using the language of economics, if the decentralized mechanism (1.40)–(1.41) is stable and convergence is fast, then it steers the system to its equilibrium – without gathering all the information to the center. This parallel has characterized the debate between the advocates and opponents in the rationality of socialist planning (Lange and Mises, see Hayek ed., 1935). Today, after the worldwide collapse of socialism, we can state that the reform proposals of Lange were quite naive and Mises' critique was correct (Kornai, 1992).

CONTROL BY INPUT AND OUTPUT STOCK SIGNALS*

The model

It is well-known that, in contrast to the capitalistic economies, in all the socialist economies input stocks were much higher than output stocks (Chikán et al., 1978; Kornai and Martos, eds, 1981b.) This peculiarity justified the introduction of maximum n^2 input stocks and the corresponding input purchases into the extended models. Then the dimension of the model jumped from n to maximum $n^2 + n$, but the model became much more realistic. It is worthwhile to outline the structure and the equations of this enlarged model.

Let $z_{i,j,t}$ and $y_{i,j,t}$ be the closing *input stocks* and *input purchases* of sector j from product i, respectively. The notations of the corresponding matrices be Z_t and Y_t, respectively. (We wrote maximum, because in a disaggregate model there are many pairs of sectors without direct connection: if $a_{ij} = 0$, then $y_{ij} = z_{ij} = 0$.)

The dynamics of the enlarged model rests on two simple identities: 1) the change in output stock is equal to output less producers' purchases less consumption; 2) the change in input stock is equal to purchase less input. The producers' inputs are proportional to outputs, the final consumption is given. The model is closed by two simple linear decentralized control rules: 1) the output of sector i is equal to the output capacity less reaction coefficient times own output stock; 2) the purchase of sector j from product i is equal to the input capacity less reaction coefficient times own input stock. We shall speak of *control by input and output stock signals*.

The equations of the model

We can now present the equations of the enlarged model.
Change in output stocks

$$(2.18) \qquad z_t = z_{t-1} + y_t - Y_t \mathbf{1} - c.$$

Change in input stocks

$$(2.19) \qquad Z_t = Z_{t-1} + Y_t - A\langle y_t \rangle.$$

Decentralized control of output

$$(2.20) \qquad\qquad y_t = y^* - \langle d \rangle z_{t-1}.$$

Decentralized control of purchases

$$(2.21) \qquad\qquad Y_t = Y^* - D \times Z_{t-1}$$

where $D > 0$ is an $n \times n$-matrix and $\times$ is the operator of multiplication of matrices by entries.

Remark. It is noteworthy that in this model there is a further decentralization within each sector: the producer only observes its output stocks and decides on its output, while the buyer of input i for sector j only observes its input stock and decides on its purchase.

Normal state and stability

The analysis of the normal state is a routine problem, thus it is left as an exercise.

Problem 2.2. Normal state. Determine the normal state of the enlarged system on the basis of Theorem 2.3.

Introducing notation $\mathbf{1}\mathbf{1}^{\mathbf{T}}$ for the dyadic product of the column vector $\mathbf{1}$ and of the row vector $\mathbf{1}^{\mathbf{T}}$, resulting in an $n \times n$ matrix with all entries being 1 and substituting (2.20)–(2.21) into (2.18)–(2.19) yield the *enlarged system of basic equations*

$$(2.22) \quad z_t = (I - \langle d \rangle)z_{t-1} + (D \times Z_{t-1})\mathbf{1} + y^* - Y^*\mathbf{1},$$
$$(2.23) \quad Z_t = A\langle \langle d \rangle z_{t-1} \rangle + (\mathbf{1}\mathbf{1}^{\mathbf{T}} - D) \times Z_{t-1} + Y^* - A\langle y^* \rangle.$$

To apply Theorem 1.16, we have to constrain the purchase reaction coefficients as well:

$$(2.24) \qquad 0 < d \le \mathbf{1} \qquad \text{and} \qquad 0 < D \le \mathbf{1}\mathbf{1}^{\mathbf{T}}.$$

It can be shown that the (state) coefficient matrix (2.22)–(2.23) is 2-cyclic [(1.51) or (A.7)], thus Theorem 1.17 implies

Theorem 2.5. *(See Kornai and Simonovits, 1981, Theorems 42 and 61.) Assume damped feedback: (2.24).*

a) The control by input and output stock signals is stable.

b) Its maximal dampening factor is achieved for $d = 1$, $D = 11^{\mathrm{T}}$ and its value is given by

$$\Phi_{1,11^{\mathrm{T}}} = \frac{1}{\sqrt{\rho(A)}}.$$

c) The feedback is (asymptotically) oscillation-free.

Remarks. 1. It is worthwhile comparing the damped optimum of the simple model and the global optimum of the enlarged model (Theorems 2.4 and 2.5). Since $\rho(A) < 1$, $\rho(A) < \sqrt{\rho(A)}$, $1/\rho(A) > 1/\sqrt{\rho(A)}$, that is, the control by output stock signals stabilizes twice as fast as the control by input and output stock signals.

2. It would be premature to prefer the simple model to the enlarged model on the basis of faster stabilization. For example, introducing sales and output expectations, the simple model (Lovell, 1962) becomes centralized, while the enlarged version (Simonovits, 1999a) remains decentralized, (see Remark after equations (2.18)–(2.21) and Section 2.3* below).

2.3* LINEAR DECENTRALIZED CONTROL WITH EXPECTATIONS

In the previous Sections we have only referred to the role of expectations in control with stock signals. We return now to the general decentralized control introduced in Section 1.4 and apply expectations to it (Simonovits, 1999a). Obviously, our general results also apply to the concrete economic models. First we recapitulate the corresponding parts of Section 1.4.

Decentralized linear control systems

Let n be a positive integer and let x and u be the n-dimensional *state* and the *control vectors*, respectively. Let B be the $n \times n$ *input matrix*. The n-vector of *perturbations* p be zero. We can now describe the time-invariant linear system's *state space equation in discrete time*:

(2.25) $x_t = x_{t-1} + Bu_t, \quad t = 1, 2, \ldots, \quad x_0$ is the initial state.

We shall need the coordinate-wise form of (2.25):

$$x_{i,t} = x_{i,t-1} + b_{ii}u_{i,t} + \sum_{j \neq i} b_{ij}u_{j,t}, \qquad i = 1, \ldots, n.$$

Let us normalize the *own-effects* as $b_{ii} = 1$ and introduce the *matrix of cross-effects*:

$$(2.26) \qquad\qquad N = I - B.$$

The diagonal entries of N are all zeros, and following Metzler (1945) we shall assume that the off-diagonal entries of N are all nonnegative: $n_{ij} \geq 0$, that is, $N \geq 0$. To exclude uninteresting cases, we shall assume that N is irreducible. A further assumption is that N is a stable matrix: $\rho(N) < 1$ (Theorem 1.4).

With the help of N, (2.25) can be rewritten as

$$(2.27) \qquad\qquad x_t = x_{t-1} + u_t - Nu_t.$$

Introducing the notion of *impact* $v_t = Nu_t$, (2.27) is replaced by

$$(2.28) \qquad\qquad x_t = x_{t-1} + u_t - v_t.$$

We have arrived at the core of the problem: the change in the state $x_{i,t}$ is decomposed into two parts: (i) the direct effect of the assigned variable $u_{i,t}$ and (ii) the impact (due to the cross-effects) $v_{i,t}$.

Expectations

Controlling the decentralized system at time t, controller i knows its own control $u_{i,t}$ but has only a forecast on impact $v_{i,t}$ of other decisions. This forecast can be called an *expectation* and denoted as $v_{i,t}^{e}$.

We shall study two types of expectations: rational expectations (or its deterministic version: perfect foresight) and naive expectations.

In the case of *rational expectations*, the forecast is equal to the actual impact:

$$(2.29) \qquad\qquad v_t^{e} = v_t.$$

In the case of *naive expectations*, the forecast is equal to the previous impact:

$$(2.30) \qquad\qquad v_t^{e} = v_{t-1}.$$

Note that at the end of time-period t, controller i knows $v_{i,t}$ from $x_{i,t} = x_{i,t-1} + u_{i,t} - v_{i,t}$, thus naive expectations are consistent with decentralization. Since at the beginning of time-period t, controller i cannot simply determine $v_{i,t}$, rational expectations requires some fast learning process which is generally not modelled (for example, Grandmont, 1998).

Remark.   A good deal of papers generalize naive expectations into adaptive expectations. In the case of *adaptive expectations* the forecast is equal to the convex combination of previous impact and previous forecast:

$$(2.31) \qquad v_t^e = \langle a \rangle v_{t-1} + \langle 1 - a \rangle v_{t-1}^e \quad \text{where} \quad 0 < a \le 1.$$

To save notations and calculations, we shall only study naive rather than adaptive expectations, when $a = 1$.

Control by expectations

Any control rule has some norm which guides the behavior of the controllers. We shall define the *normal state* x_t^e to be proportional to the expected impact v_t^e where the proportionality vector h to be called the *norm vector* is given:

$$x_{i,t}^e = h_i v_{i,t}^e, \qquad i = 1, \ldots, n;$$

or in vector form:

$$(2.32) \qquad\qquad\qquad x_t^e = \langle h \rangle v_t^e.$$

Note that here the normal state is different from the steady state. It would be rather costly to steer the system to the normal state in one step, thus the control aims at the *planned state* x_t^P. This latter is a convex combination of the normal state x_t^e and the previous state x_{t-1}:

$$(2.33) \qquad x_t^P = \langle k \rangle x_t^e + \langle 1 - k \rangle x_{t-1}, \quad 0 < k \le 1.$$

Vector k is called *reaction vector*.

By the definition of planned state, control u_t is determined from (2.28), by replacing the new state by the planned state and the impact by its expected value: $x_t^{\mathrm{p}} = x_{t-1} + u_t - v_t^{\mathrm{e}}$ or in view of (2.32)–(2.33), $u_t = (I + \langle k \rangle \langle h \rangle) v_t^{\mathrm{e}} - \langle k \rangle x_{t-1}$.

Introducing notation $\langle b \rangle = I + \langle k \rangle \langle h \rangle$, the control is given by

$$(2.34) \qquad\qquad u_t = \langle b \rangle v_t^{\mathrm{e}} - \langle k \rangle x_{t-1}.$$

Substituting (2.34) into $v_t = N u_t$, the impact is expressed as a function of the expectations and the past state:

$$(2.35) \qquad\qquad v_t = N \langle b \rangle v_t^{\mathrm{e}} - N \langle k \rangle x_{t-1}.$$

By successive elimination, it is possible to obtain the standard form $x_t = M x_{t-1}$, but the results would be awkward. Rather we choose the characteristic root technique. To do so, we shall need another relation. Substituting (2.34) into (2.28) yields

$$(2.36) \qquad\qquad v_t + x_t = \langle \mathbf{1} - k \rangle x_{t-1} + \langle b \rangle v_t^{\mathrm{e}}.$$

Leaving aside the definitions of steady state, initial conditions and its stability, we shall analyze the model and present the results. Following the order of difficulty, we shall start the discussion with rational expectations and then continue with naive expectations.

Rational expectations

According to the main-stream approach, rational expectations is *the* suitable assumption on expectations.

Substituting (2.29) into (2.35), yields $v_t = N \langle b \rangle v_t - N \langle k \rangle x_{t-1}$, hence

$$(2.37) \qquad\qquad (I - N \langle b \rangle) v_t = -N \langle k \rangle x_{t-1}.$$

Note that (2.37) is an implicit equation, which may have no solution, one solution or more than one solution. To avoid unnecessary complications with (2.37), let us assume that

$$(2.38) \qquad\qquad \rho(N \langle \mathbf{1} + h \rangle) < 1.$$

Since $\rho(N) < 1$ and the spectral radius of a nonnegative irreducible matrix is an increasing function of its elements (Theorem A.7d), (2.38) amounts to the assumption that the norm vector is sufficiently small.

Since $0 < k \leq \mathbf{1}$, $b \leq \mathbf{1} + h$, thus (2.38) and the monotonicity rule imply the existence and positivity of the Leontief-inverse of $N \langle b \rangle$, $(I - N \langle b \rangle)^{-1}$, hence (2.37) reduces to

$$(2.39) \qquad\qquad v_t^{\mathrm{e}} = v_t = -(I - N \langle b \rangle)^{-1} N \langle k \rangle x_{t-1}.$$

Thus for rational expectations, the impact vector is a linear function of the past state vector. We shall prove

Theorem 2.6. *Assume (2.38) holds. a) For the norm vector h, the control by rational expectations is stable if and only if the interactions represented by N and the reactions are sufficiently weak:*

$$(2.40) \qquad \rho(N\langle 2b - k\rangle\langle 2 \cdot \mathbf{1} - k\rangle^{-1}) < 1.$$

b) Fix the norm vector. If the control is stable for a reaction vector, then it is also stable for any weaker reaction, represented by k' with $0 < k' \le k$.

c) For the norm vector h, the control by rational expectations is stable for any reaction if and only if the norms are sufficiently small:

$$(2.41) \qquad \rho(N\langle \mathbf{1} + 2h\rangle) < 1.$$

Remarks. 1. Assertion b) appears to be natural, but it is not true in other decentralized control models (Simonovits, 1981a). Assertion c) is a direct consequence of a) and b).

2. Concerning the enlarged control model with stock signalling, Theorem 2.5 demonstrated that $\rho(N) = \sqrt{\rho(A)}$. In open economies, $\rho(A)$ is about $1/2$, thus (2.41) yields quite a large room for the norms.

Proof. Let λ be an arbitrary characteristic root of M in the standard form. Then the corresponding basic solution of the homogeneous difference equation is

$$(2.42) \qquad x_{t-1} = \lambda^t x_0, \quad u_t = \lambda^t u_0 \quad \text{and} \quad v_t = \lambda^t v_0.$$

Following the idea of Lemma 1.1 after Theorem 1.16, the problem is transformed into a nonlinear fixed-point problem.

Substitute (2.42) into (2.36): $v + \lambda x = \langle \mathbf{1} - k\rangle x + \langle b\rangle v$, that is, $x = \langle(\lambda - 1)\mathbf{1} + k\rangle^{-1}\langle b - \mathbf{1}\rangle v$. Inserting our last expression into (2.35) leads to $v = N\langle b\rangle v - N\langle k\rangle\langle(\lambda - 1)\mathbf{1} + k\rangle^{-1}\langle b - \mathbf{1}\rangle v$, that is,

$$(2.43) \qquad v = N\langle p(\lambda)\rangle v$$

where

$$(2.44) \qquad \langle p(\lambda)\rangle = \langle(\lambda - 1)b + k\rangle\langle(\lambda - 1)\mathbf{1} + k\rangle^{-1}.$$

Before going further, we mention the following

Problem 2.3. Let λ be complex number, α, β and ε be positive numbers satisfying $\varepsilon < 1$ and $\varepsilon\beta < \alpha < \beta$. Let

$$\pi(\lambda) = \frac{\beta\lambda - \alpha}{\lambda - \varepsilon}.$$

Then $|\pi(\lambda)| < \pi(-1)$ if $|\lambda| = 1$ and $\lambda \neq -1$.

Continuation of the proof. We shall prove that at the stability frontier $|\lambda| = 1$, $\lambda = -1$ holds, yielding (2.40).

The proof is by contradiction and is a variant of the well-known proof of the generalization of the monotonicity rule (Theorem A.7d): let G be an arbitrary matrix and let $G^{\#} = (|G_{ij}|)$ be the $n \times n$ matrix formed by its moduli, then $\rho(G) \leq \rho(G^{\#})$.

Take the moduli of the expressions occurring in (2.43): $v^{\#} < N\langle p(\lambda)^{\#}\rangle v^{\#}$. Letting $\beta_i = b_i$, $\alpha_i = b_i - k_i$ and $\varepsilon_i = 1 - k_i$, $\pi_i = p_i$, $i = 1,\ldots,n$, Problem 2.3 implies that $p(\lambda_1)^{\#} < p(-1)$ for $|\lambda_1| = 1$, $\lambda_1 \neq -1$. Relying on the monotonicity rule, we arrive at $1 < \rho(N\langle p(\lambda_1)^{\#}\rangle) < \rho(N\langle p(-1)\rangle)$.

Since $\lim_{\lambda \to -\infty} \rho(N\langle p(\lambda)\rangle) = 0$, there exists a scalar λ^*, $-\infty < \lambda^* < -1$, such that $\rho(N\langle p(\lambda^*)\rangle) = 1$. Since $|\lambda^*| > 1$, λ is not a dominant characteristic root. Contradiction. ∎

Problem 2.4. Prove (2.40) for the special case $h = \chi\mathbf{1}$ and $k = \kappa\mathbf{1}$.

Naive expectations

Having experienced some definitional problems with rational expectations in (2.37) and taking into account Lovell's (1986) critique mentioned in the Introduction, it is worthwhile considering other expectations. The simplest alternative is naive expectations given by (2.30). Modifying the procedure above, we obtain

Theorem 2.7. a) *For the norm vector h, the control by naive expectations is stable if the interactions represented by N and the reactions are sufficiently weak:*

$$(2.45) \qquad \rho(N\langle 2b + k\rangle\langle 2 \cdot \mathbf{1} - k\rangle^{-1}) < 1.$$

If matrix N is 2-cyclic, then (2.45) is not only a sufficient but a necessary condition as well.

b) Fix the norm vector. If the control is stable for a reaction vector k, then it is also stable for any k' such that $0 < k' \leq k$.

c) For the norm vector h, the control by naive expectations is stable for any reaction if and only if both the interactions represented by N and the norms are sufficiently small:

(2.46) $$\rho(N\langle 3 \cdot \mathbf{1} + 2b\rangle) < 1.$$

d) Assume again (2.38). If the economy is stable with naive expectations, then it is also stable with rational expectations.

Remarks.   1. Assertion d) appears as natural, but it is not true either in Lovell (1962) or Simonovits (1982).

2. It is remarkable that while the mechanism of naive expectations corresponds to the Gauss–Seidel iterative method of solution of linear equations, the mechanism of rational expectations is a counterpart of block-relaxation (Young, 1971). For nonnegative matrices, the second method is faster than the first in numerical analysis, too.

3. For very low norms, (2.46) reduces to $\rho(N) < 1/3$, which gives a quite tight but not absurd upper bound in the extended stock signalling model.

Problem 2.5. Prove Theorem 2.7. Hint: Because v_t is replaced v_{t-1} at the appropriate places, (2.44) is replaced by

$$\langle p(\lambda)\rangle = \langle(1 - \lambda^{-1})b + k\rangle\langle(\lambda - 1)\mathbf{1} + k\rangle^{-1}.$$

3. NONLINEAR DIFFERENCE EQUATIONS

In this Chapter we shall outline those elements of the theory of *time-invariant nonlinear difference equations* which will be used in the book. In Section 3.1 we shall investigate the existence and stability of *fixed points*, extending theorems for linear systems to nonlinear ones. In Section 3.2 we shall study *limit cycles*. In Section 3.3 the so-called *chaotic dynamics* will be analyzed where the paths depend sensitively on the initial states. Useful information can be found in Guckenheimer (1979), Guckenheimer and Holmes (1986), Grandmont (1986) and Devaney (1989).

3.1 EXISTENCE AND STABILITY OF FIXED POINTS

We return to the general time-invariant difference equations introduced in Section 1.1. We concentrate on the existence and stability of a fixed point of a system of nonlinear difference equations. We repeat the formula of vector difference equation:

$$(3.1) \qquad x_t = f(x_{t-1}), \qquad t = 1, 2 \ldots, \qquad x_0 \quad \text{given;}$$

and that of its fixed point:

$$x^\circ = f(x^\circ).$$

Existence of a fixed point

We shall first consider the existence of a fixed point.

Theorem 3.1. *(Brouwer fixed-point theorem, Arrow and Hahn, 1971, Appendix C, Theorem 1.) If a continuous function f maps a compact and convex set $\mathcal{X}$ of $\mathbf{R}^n$ into itself (invariance), then it has at least one fixed point.*

Remarks.   1. This theorem is of fundamental importance in both mathematics and economics. On the one hand, it expresses an essential property of continuous functions mapping n-dimensional domains into themselves. On the other hand, this theorem is the foundation of the general equilibrium theory.

2. Nevertheless, our theorem does not show us how to compute the fixed point. Scarf (1967) gave an algorithm for finding an approximate fixed point: for any positive ε, it provides a point x^ε for which $|x^\varepsilon - f(x^\varepsilon)| < \varepsilon$. However, it is conceivable that x^ε lies far away from any exact fixed point x°.

The role of the assumption of compactness is obvious, the role of convexity will be shown by

Example 3.1. Nonconvexity $\Rightarrow$ no fixed point. Let $\mathcal{X}$ be the annulus $1 \leq x_1^2 + x_2^2 \leq 4$ and f be the rotation by angle $\pi/2$ around the origin. Then $\mathcal{X}$ is compact but not convex, function f is continuous but it has no fixed point. (For the disk $x_1^2 + x_2^2 \leq 4$, f would have a fixed point: $0 = f(0)$.)

The following example shows a well-known special case of the fixed-point theorem.

Example 3.2. Crossing property. For a scalar function $(n = 1)$, Theorem 3.1 reduces to the well-known Bolzano-theorem: in a compact interval, every continuous function takes on every intermediate value. Indeed, $\mathcal{X} = [a, b]$, $f(b) - b < 0 < f(a) - a$, that is, there exists an intermediate point $x^\circ \in \mathcal{X}$ such that $f(x^\circ) - x^\circ = 0$.

No dynamic analysis can stop at the study of fixed points, it should also enquire whether the system, starting out of a fixed point, has a tendency to return to it.

Global stability

In the case of global stability we have to prove convergence for an arbitrary initial state. For the time being, we assume that there exists at least one fixed point. Assume that $f : \mathcal{X} \to \mathcal{X}$ where $\mathcal{X}$ is a compact set in $\mathbf{R}^n$.

It was Lyapunov's pioneering idea that one can study the stability of a dynamical system without having an explicit solution. What we only need is to have a suitable positive function, measuring the 'distance' from the fixed point, and the distance is decreasing in time. Formally: a continuous function $V : \mathcal{X} \to \mathbf{R}$ is called a *Lyapunov function* with respect to function f if

(i) it is zero at the fixed point and positive otherwise:

$$V(x^\circ) = 0 \quad \text{and} \quad V(x) > 0 \quad (x \neq x^\circ);$$

(ii) for almost any pair of states $(x, f(x))$, V is decreasing:

$$V[f(x)] < V(x) \quad \text{unless} \quad x = x^\circ.$$

Theorem 3.2. *(Lyapunov, 1893.) If there exists a Lyapunov function with respect to the dynamics, then the fixed point is globally stable.*

Remarks.   1. In general it is quite difficult to find Lyapunov functions. For special functions, however, there are methods to find them. For example, for every stable linear system of difference equations, there exists a quadratic Lyapunov function. To give an example, consider the case (the spiral) described in Problem 1.8 where $V(x) = |x|$ is a Lyapunov function. It is decreasing for $\rho < 1$ and increasing for $\rho > 1$, implying stability and instability, respectively.

2. It is worthwhile to mention the physical background to this theorem: in a dissipative system (where part of the mechanical energy is transformed into other types of energy), the mechanical energy diminishes until the system reaches its equilibrium.

Proof.   Let $x_0 \in \mathcal{X}$ be an arbitrary initial state and let $\{x_t\}$ be the emerging path. If the path jumps to the fixed point, stability is trivial, thus we can exclude this degenerate case. Then sequence $\{V(x_t)\}$ is positive and decreasing, it has a limit point, denoted by V^*. Since $\mathcal{X}$ is compact, it has a convergent subsequence $\{x_{t_j}\}$ with a limit x^*. Because V is continuous, $V(x^*) = V^*$ and $\{f(x_{t_j})\}$ converges to $f(x^*)$. But $\{V(x_t)\}$ is convergent, thus the two limits are equal: $V(x^*) = V[f(x^*)]$. By (ii), $x^* = x^\circ$. If any subsequence converges to the same limit, then the sequence is also convergent.   ∎

Local stability

In contrast to global stability, local stability only guarantees convergence for small enough neighborhood of a fixed point. Let f have a continuous derivative at its fixed point x°, and let

$$M = \mathbf{D}f(x^\circ), \qquad w = f(x^\circ) - Mx^\circ$$

where $\mathbf{D}$ is the differential operator. Now

$$x_t = Mx_{t-1} + w$$

is called the *linearized part* of (3.1).

We present the generalization of Theorem 1.4:

Theorem 3.3. *The discrete-time nonlinear system (3.1) is locally stable if its linearized part is stable, that is, if the spectral radius is less than unity:*

$$\rho(M) < 1.$$

Remarks. 1. Obviously, for any fixed time-interval, the solution of a vector difference equation hardly changes when the R.H.S. is changed a little. Theorem 3.3 extends this type of continuity for the entire time-axis.

2. Similarly, instability is implied by $\rho(M) > 1$. Nothing can be said, however, in the neutral case where $\rho(M) = 1$, analogously to the collapse of the second-order condition of maximum for zero second derivative.

Sketch of the proof. We shall discuss the continuous-time version of Theorem 3.3 in Section 5.4 and allude to the proof. Here we only underline the essence of the proof for discrete time. In case of a stable linearized part, one can construct a Lyapunov function. Hence stability is established.

For the scalar case, the proof goes like this. Let $x^\circ = 0$, then $f(x) = Mx + \vartheta(x)$ where $\lim_{x \to 0} \vartheta(x)/x = 0$. Our candidate Lyapunov function is $V(x) = x^2$. Then the nonnegativity condition is satisfied and the monotonicity condition $V[f(x)] < V(x)$, that is $M^2x^2 + 2Mx\vartheta(x) + \vartheta(x)^2 < x^2$ holds if $\varepsilon = |\vartheta(x)/x|$ and

$$\varepsilon < \min \left[\frac{\sqrt{1 - M^2}}{2}, \frac{1 - M^2}{4|M|} \right].$$

It is enough if $|x|$ is small enough. ∎

In the remaining part of the Chapter a prominent role will be played by the simplest nonlinear scalar function and the corresponding equation:

Example 3.3. Logistic function. Let $n = 1$ and

$$f(x) = ax(1-x), \quad 0 \le x \le 1 \quad \text{and} \quad 0 \le a \le 4.$$

Fixed points: $x^\circ = 1 - 1/a$ and $x^\circ = 0$. Since the latter is unstable ($f'(0) = a > 1$), we shall only concentrate on the former. Viability: it is important that the logistic map transform the unit interval into itself. $0 \le f(x) \le f(1/2) = a/4$ implies $0 \le a \le 4$.

Stability: $f'(x) = a(1 - 2x)$ implies $f'(x^\circ) = 2 - a$; stability region: $1 < a < 3$.

Global stability again

We return now to the global stability. Moreover, the existence and the uniqueness of the fixed point are proved rather than assumed, but in turn, quite strong assumption is made on function f.

We shall need the following notion. A function $f : \mathbf{R}^n \to \mathbf{R}^n$, defined on a compact set $\mathcal{X}$, is called a *contraction* if $f(\mathcal{X}) \subseteq \mathcal{X}$ and the distance between two images of two different points is smaller than the distance between the two points:

(3.2) $\quad ||f(x) - f(y)|| < ||x - y|| \quad \text{for every} \quad x, y \in \mathcal{X}, \quad x \ne y.$

Remark. Because of compactness of $\mathcal{X}$ and the continuity of the norm, (3.2) can be sharpened as follows: there exists a real δ, $0 < \delta < 1$ such that

$$||f(x) - f(y)|| < \delta ||x - y|| \quad \text{for every} \quad x, y \in \mathcal{X}.$$

If contraction is defined by this formula, then there is no need on the assumption of compactness.

We can now formulate

Theorem 3.4. *Banach fixed-point theorem. If f is a contraction on a compact set $\mathcal{X}$, then (3.1) has a unique fixed point which is globally stable.*

Remark.* This theorem holds for much more general spaces than $\mathbf{R}^n$. In that setting the points of the space are not finite-dimensional vectors but abstract objects, for example, continuous functions mentioned after Problem A.5; the members of the sequence are the successive approximate solutions of a differential equation, and the limit is the exact solution (Theorem 5.2).

Proof. Using mathematical induction and the stronger definition of contraction, we have

$$||x_{t+1} - x_t|| = ||f(x_t) - f(x_{t-1})|| < \delta||x_t - x_{t-1}|| < \delta^t||x_1 - x_0||.$$

Hence, applying the triangle inequality repeatedly for integers $u > t$,

$$||x_u - x_t|| \leq ||x_u - x_{u-1}|| + \cdots + ||x_{t+1} - x_t||$$

$$< \sum_{k=t}^{u-1} \delta^k ||x_1 - x_0|| \leq \frac{\delta^t}{1-\delta}||x_1 - x_0||.$$

Therefore $\{x_t\}$ is a Cauchy sequence, which has a limit in $\mathcal{X}$. $x^\circ = \lim_t x_t = \lim_t f(x_{t-1}) = f(x^\circ)$, that is, x° is a fixed point.

To prove uniqueness, assume we have two fixed points: $y^\circ \neq x^\circ$. Then $||x^\circ - y^\circ|| = ||f(x^\circ) - f(y^\circ)|| < ||x^\circ - y^\circ||$, contradiction. ∎

Remarks. 1. For $n = 1$, function f is a contraction on a compact interval I if $|f'(x)| < 1$ for every $x \in I$. Indeed, by Lagrange's intermediate value theorem, for any x and y, $x < y$, there exists a z, $x < z < y$ such that $f(x) - f(y) = f'(z)(x-y)$, that is, $|f(x) - f(y)| < |x - y|$.

2. For $n > 1$, the condition of contraction is much more involved, even in the linear case $f(x) = Mx + w$ (Halmos, 1958 and Young, 1971, Section 3.7). Taking into account the information on matrix norms (Appendix A), it is clear that stability of the matrix $[\rho(M) < 1]$ is a necessary, while $||M|| < 1$ or even $||M^k|| < 1$ is a sufficient condition (Theorem 1.8).

3. Note that now $V(x) = ||x - x^\circ||$ is a Lyapunov function.

The following problems will illustrate the limits of applicability of the contraction theorem.

Problem 3.1. Lack of fixed point. a) Demonstrate that $f(x) = x + (1+e^x)^{-1}$ has no fixed point in $I = (-\infty, \infty)$ though $0 < f'(x) < 1$ holds for $x \in I$. b) Why is Theorem 3.4 not applicable?

For scalar functions, the *Lamerey stair* is a very useful graphic tool (Arnold, 1973). It is a collection of horizontal and vertical lines connecting points $(x_t, f(x_t))$ and (x_t, x_t) for $t = 1, 2, \ldots$. For example, the image of point x_0 is $f(x_0)$: a vertical line leads from $(x_0, 0)$ to $(x_0, f(x_0))$. Let $x_1 = f(x_0)$. Hence a horizontal line cuts point (x_1, x_1) from line $y = x$. Figures 3.1 and 3.2 display this stair for two logistic functions. (Economists know it from the so-called cobweb model discussed by Ezekiel, 1938, see also Hommes, 1994.)

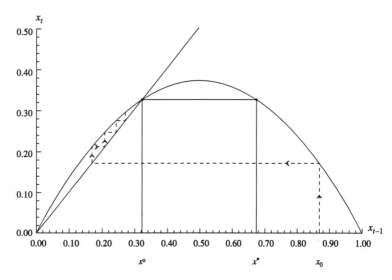

Figure 3.1
Globally stable fixed point: $a = 1.5$

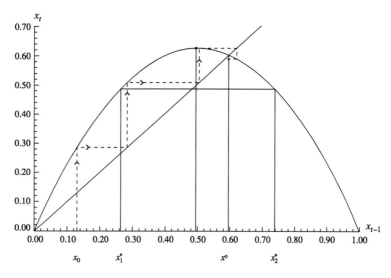

Figure 3.2
Globally stable fixed point: a = 2.5

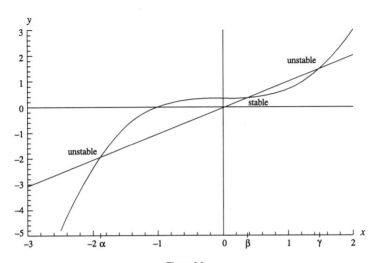

Figure 3.3
Locally stable, globally unstable fixed point

Problem 3.2. Finding a square root. A pocket calculator computes the square root of a positive number β by the following iterative method:

$$x_t = \frac{1}{2}\left(x_{t-1} + \frac{\beta}{x_{t-1}}\right)$$

where x_0 is an arbitrary positive number.

a) Prove convergence by the Lamerey stair for $\beta < x^\circ < \infty$.

b) Demonstrate that for small enough x's, the function is not a contraction but the algorithm converges to $\sqrt{\beta}$.

Problem 3.3.* Global stability for the logistic equation (Devaney, 1989). a) Prove that for the logistic equation, *local stability implies global one* $(1 < a < 3)$. (Hint: Separate the two cases: $1 < a < 2$ and $2 < a < 3$ and check what happens during the first few iterations.) b) What is the dampening factor for $a = 2$?

We now show an example where local stability does *not* imply global stability.

Example 3.4. Coexistence of three fixed points. Let $f(x) = (x^3 + 1)/3$. a) By Figure 3.3, there are three fixed points, their notations are α, β, γ, respectively where $-2 < \alpha < -1 < 0 < \beta < 1 < \gamma < 2$. b) There is only one locally stable fixed point, namely, β, with a *domain of attraction* being the open interval (α, γ). For initial points from $(-\infty, \alpha)$ and (γ, ∞), the paths go to infinity.

a) A simple calculation suffices to check that for $p(x) = f(x) - x$, the following inequalities hold: $p(-2) < 0$, $p(-1) > 0$, $p(0) > 0$, $p(1) < 0$, $p(2) > 0$. Hence the existence and the location of α, β, γ are established. b) $f'(x) = x^2$, thus f is a contraction exactly in interval $(-1,1)$. The stability of paths starting from the remaining intervals $(\alpha, -1)$ and $(1, \gamma)$ can be proved directly: because of $f'(x) > 1$, $x_0 \in (\alpha, -1)$ implies $\alpha < x_0 < f(x_0) < 0$, therefore the generated sequence sooner or later enters $(-1,1)$. Note that using higher degree polynomials or other appropriate functions, one can get more than one locally stable fixed point.

3.2 LIMIT CYCLES

We repeat first the definition of P-cycles from Section 1.1. Let P be an integer larger than 1. A series of vectors $x_1, x_2, \ldots, x_P$ is called a *P-cycle* of system f if a path starting from x_1 goes through $x_2, \ldots, x_P$ and returns to x_1.

Not only a fixed point, but a cycle also can be stable. Cycle $x_1, x_2, \ldots, x_P$ is called a *(local) limit P-cycle* of system f if any initial state y_1 close enough to x_1 generates a y-path, which converges to the P-cycle:

$$(3.3) \qquad \text{If} \quad y_1 \quad \text{is close to} \quad x_1, \quad \text{then} \quad \lim_{k \to \infty} y_{kP+Q} = x_Q$$

where k and Q are integers, $Q = 1, \ldots, P$.

Remarks. 1. Notice that cycle $x_1, x_2, \ldots, x_P$ is identical to cycle $x_2, \ldots, x_P, x_1$. Therefore at the formulation of (3.3) we have to be careful.

2. As is usual, we defined a *locally* stable cycle. If we want to stress that any initial state (except for few initial states, like the unstable stationary state) can be considered, we use the adjective *global*.

3. In an alternative way (for example, Devaney, 1989) one can define *periodic points* by the return property. To understand the similarity between cycles and fixed points, we shall introduce the *iterations*

of function f:
$$f^1(x) = f(x), \ldots, f^t(x) = f(f^{t-1}(x)), \qquad t = 2, 3, \ldots.$$
Note that $x_t = f^t(x_0)$.

Theorem 3.5. *Cycle and fixed point.*

a) $x_1, x_2, \ldots, x_P$ is a cycle of (3.1) with period P if and only if x_Q is a nontrivial fixed point of the P-iteration
$$z_t = f^P(z_{t-1}),$$
that is,
$$x_Q = f^P(x_Q) \neq f(x_Q), \qquad Q = 1, 2, \ldots, P.$$

b) *The cycle in a) is a limit cycle if and only if the corresponding fixed points of f^P are stable, or equivalently, the corresponding matrix is stable:*
$$(3.4) \qquad \rho[\mathbf{D}f(x_P) \cdots \mathbf{D}f(x_1)] < 1.$$

Proof. a) By the definition of f^P.

b) The first part is obvious. By Theorem 3.3, stability follows from $\rho[\mathbf{D}f^P(x_1)] < 1$ and by the chain rule, $\mathbf{D}f^P(x_1) = \mathbf{D}f(x_P) \cdots \mathbf{D}f(x_1)$. ∎

The following problem is connected to the fact that every P-cycle has P different forms.

Problem 3.4.* What happens to (3.4) if the cycle is rewritten into $x_2, \ldots, x_P, x_1$? (Hint: Prove and use the following theorem of linear algebra: matrices AB and BA have identical characteristic roots, Young, 1971, 2.1.11.)

We shall present a limit 2-cycle.

Example 3.5. Limit 2-cycle for the logistic equation. According to Theorem 3.5, we have to study
$$f^2(x) = af(x)[1 - f(x)] = a^2 x(1 - x)(1 - ax + ax^2), \quad 0 < x < 1.$$
2-cycle: $x_Q = f^2(x_Q)$, $Q = 1, 2$ and $x_Q \neq x^\circ$. We obtain a cubic equation $a^3 x^3 - 2a^3 x^2 + a^2(1+a)x + 1 - a^2 = 0$. Having divided it by $a(x - x^\circ)$, we obtain the quadratic equation $a^2 x^2 - a(a+1)x + (a+1) = 0$ which has two real roots:
$$x_{1,2} = \frac{(a+1) \pm \sqrt{(a+1)(a-3)}}{2a},$$
both lying between 0 and 1 if and only if $3 < a < 4$.

Stability: By (3.4), $f^{2'}(x_1) = f'(x_1)f'(x_2) = -a^2 + 2a + 4$; $|f^{2'}(x_1)| < 1$, that is, $3 < a < 1 + \sqrt{6} \approx 3.44$. ∎

Remark. We shall see in Theorem 3.8 that for larger a's, limit cycles with periods 4, 8, 16, and so on occur (period-doubling, or bifurcation), with components appearing in algebraic equations with degrees 4, 8, 16, and so on. Probably one cannot solve them explicitly except for $P = 4$. Thus we are forced to use a computer even at this early stage. (In practice already the calculation of square root is done on a computer, see Problem 3.2.)

Attractor and quasi-cycle*

After reviewing (stable) fixed points and (limit) cycles we should discuss the quasi-cycle, the linear version of which we mentioned after Theorem 1.10. To do so, however, we would need sophisticated topological tools. Here we only outline the definition together with that of attractor. (Exact definitions are to be found in Devaney, 1989.) A set $J \subseteq \mathbf{R}^n$ is called an *invariant set* of the dynamical system if $f(J) \subseteq J$. A fixed point or a cycle is an invariant set but there are more sophisticated invariant sets, too.

Loosely speaking, set $\mathcal{A}$ is called an *attractor* of a system f if for any point $a \in \mathcal{A}$, there exists a suitable initial state $x_a \in \mathcal{X}$ such that a is a *limit point* of the sequence $\{f^t(x_a)\}_t$. Roughly speaking, the dynamical system f is said to have a *quasi-cyclic* attractor if f, confined to the attractor, is 'similar' to a circle rotation with an irrational rotation number.

3.3 CHAOS

From now on we shall study such nonlinear systems which are neither stable, nor cyclic nor quasi-cyclic, they are chaotic. (For a more complete description, see for example, Devaney, 1989.)

Insensitive dependence on the initial state

It may appear obvious that a path of a dynamical system hardly changes if the initial state is slightly perturbed. We shall say that the dynamical system (3.1) is *insensitive to the initial state* x_0 if

a) the paths are bounded and b) initial states close to x_0 generate paths close to $\{x_t\}$. Formally, for every $\varepsilon > 0$, there exists a $\delta_\varepsilon > 0$ such that if $||x_0 - y_0|| < \delta_\varepsilon$, then $||x_t - y_t|| < \varepsilon$ for all t's.

Laplacian determinism implicitly assumed this feature when claimed the following: the initial state determines the entire path of a dynamical system. Adams and Leverrier's forecast of the Neptune in 1846 was a most notable example of the success of this principle (although luck was also needed).

We shall make several remarks.

1. It is easily seen that a bounded linear system is insensitive to any initial state. Indeed, because of boundedness, we have to consider only stable basic solutions in (1.21). Then the deviations can be kept arbitrarily small.

2. Without boundedness, many linear systems would display sensitivity to the initial state. In fact, let $x_{t+1} = 2x_t$. Then for a small δ, $x_0 = 0$ and $x_0 = \delta$ are close to each other, but $2^t\delta$ and 0 are far apart for large t's.

3. As a reminder, let us repeat that for *few* initial states even well-behaved dynamical systems are sensitive. For example, suppose a system with two attractors: a fixed point and a global limit cycle (Example 3.5). Any path starting from a point close to the fixed point converges to the limit cycle but the fixed point rests in its place.

Chaos

Until now we have concentrated on purely classical concepts which have no direct relation with chaos. We now turn to the central concept of this Section. We shall say that the dynamical system (3.1) depends *sensitively on the initial state* x_0 if a) the paths are bounded and b) there are initial states, arbitrarily close to x_0 which generate paths diverging from $\{x_t\}$. Formally, there exists an $\varepsilon > 0$ such that for every $\delta > 0$, there exist a positive scalar t_δ and a state vector y_0 such that $||x_0 - y_0|| < \delta$ and $||x_{t_\delta} - y_{t_\delta}|| > \varepsilon$.

A dynamical system f is called *truly chaotic* (in the sense of Guckenheimer, 1979) if it has *sensitive dependence on the initial values* with a positive probability. According to an earlier observation, only nonlinear functions can generate chaos.

A truly chaotic dynamical system behaves erratically at least for a set of initial states having a positive measure. This means that the Laplacian determinism does not apply to such a system, notwith-

standing the extreme simplicity and the low degree of freedom of the
system.

There are numerous other definitions of chaos. We shall only
outline just another one. A dynamical system f is called *topologically
chaotic* (in the sense of Li and Yorke, 1975) if (i) there exist infinitely
many cycles with different periods, (ii) there exists an uncountable
set of initial states generating aperiodic paths and (iii) none of the
aperiodic paths stays close to cycles for ever.

There are topologically chaotic systems that are not truly chaotic
(see Example 3.6 below). They behave erratically in the long run for
exceptional initial states, but asymptotically behave in a nice way
for almost all initial states. (Observe, however, that with positive
probability the initial states generate *transient chaos*.)

We shall make the following distinction, regardless of truly or
topological chaos: (a) A chaotic dynamical system is called *acyclic*
(or *primitive*) if it is chaotic and its attractor (determining its long
run behavior) is a connected set. (b) A chaotic dynamical system f
is called *P-cyclic* if it is chaotic and the P-iterated dynamical system
f^P displays acyclic chaos.

Strange behavior is obvious for acyclic chaos, but the uncertainty
may be weaker for cyclic chaos where it is at least known that the
system jumps from the connected set $\mathcal{A}_Q$ to another connected set
$\mathcal{A}_{Q+1}$ asymptotically, $Q = 1, 2, \ldots, P$ where $\mathcal{A}_{P+1} = \mathcal{A}_1$ (Hommes,
1991, Chapter 2A).

It is possible that the ranges of the variables (sets $\mathcal{A}_Q$) are so
small that the qualitatively erratic dynamics can quantitatively be
conceived as a fixed point (if $P = 1$) or a limit cycle (if $P > 1$). In
these cases the mathematical tools are too sharp to be used *directly*
in economics.

We now present the *tent map* which illustrates (in Figure 3.4) the
behavior of the logistic function for $a = 4$:

$$(3.5) \qquad f(x) = \begin{cases} 2x & \text{if } 0 \leq x \leq 1/2; \\ 2 - 2x & \text{if } 1/2 < x \leq 1. \end{cases}$$

In words: the function increases linearly in the first interval and
decreases linearly in the second.

As a starter, consider

Problem 3.5. Show that a) the genuine fixed point of the tent
map is $x^\circ = 2/3$; b) its 2-cycle is $x_1 = 2/5$ and $x_2 = 4/5$, c) it has

two 3-cycles, namely $\{2/9, 4/9, 8/9\}$ and $\{2/7, 4/7, 6/7\}$; d) all are unstable.

Theorem 3.6. *(Ito et al., 1979.) The tent map (3.5) is chaotic.*

Instead of proof. Evidently the tth iterate of f goes up linearly on 2^{t-1} equally spaced intervals and down on another 2^{t-1} equally spaced intervals, in the range [0,1] one after the other. The number of fixed points is equal to 2^t, of which 2^{k-1} fixed points yield the 2^{k-1}-cycles appearing for the kth iteration, $k = 1, 2, \ldots, t$. The sensitive dependence on the initial state is obvious here: regardless of the closeness of initial states x_0 and y_0, there exists a separating dyadic division point $s/2^t$ (s and t being integers), that is, after iteration t x_u and y_u sit on different branches. ∎

What can we say of the limits of iterations if we consider a general scalar continuous unimodal function (function with one hump)?

Theorem 3.7. *Let f be a continuous unimodal function which maps interval $I = [a, b]$ into itself. Assume f has a 3-cycle, that is, there exists a point c, such that*

$$f(c) \neq f^2(c) \neq f^3(c) = c.$$

a) *(Sharkovskii, 1964.) Then for any positive integer $P > 1$, the dynamical system has a P-cycle.*

b) *(Li and Yorke, 1975.) Then the dynamical system is topologically chaotic.*

Remarks. 1. The proof is elementary but rather lengthy and involved.

2. Sharkovskii (1964) proved a little more than a): there exists a universal order (independently of f) of positive integers except 1, say such that if P precedes Q in this order and f has a Q-cycle, then it has a P-cycle, too. Rather than presenting this order we mention that 2 is the first and 3 is the last element.

3. It is easy to see that many functions have 3-cycles (for example, Problem 3.5), but it is quite surprising that all such functions have P-cycles for any P. Imagine a double sequence $\{x_{1,P}, x_{2,P}, \ldots, x_{P,P}\}_{P=2}^{\infty}$ in interval (0,1) such that $f : x_{1,P} \to x_{2,P} \to \ldots \to x_{P,P} \to x_{1,P}$ for $P = 2, 3, \ldots$. Small wonder that such a system behaves very wildly as Li and Yorke state. Or does it?

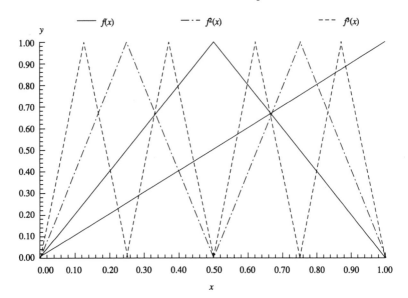

Figure 3.4
Tent map and its iterates

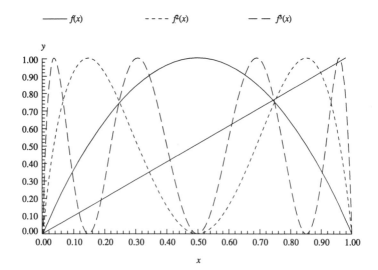

Figure 3.5
Logistic map and its iterates

Computer simulations are widely used in this field. They help in deducing the properties of certain functions to be used in analytical studies and substitute the missing analysis. An additional advantage of computer simulations is that they highlight certain quantitative results. First we simply study the paths of a dynamical system.

Problem 3.6. With the help of transformation $x_t = \sin^2 \varphi_t$, prove that $x_t = 4x_{t-1}(1 - x_{t-1})$ is chaotic. Figure 3.5 displays the original function and its 2nd, 3rd iterates. Note the similarity with Figure 3.4.

Problem 3.7. Paths of logistic equation. Write a computer program for studying the paths of logistic equation for the following values of parameter: $a = 1; 2; 3; 3.5; 3.839, 4$ and different initial values.

Example 3.6. One or infinite cycles. (Devaney, 1989, Chapter 1.13.) Let $a = 3.839$. a) Then the sequence $x_{1,3} = 0.149888$, $x_{2,3} = 0.489172$ and $x_{3,3} = 0.959299$ is a globally stable 3-cycle. b) Consequently, all the other P-cycles $(P = 2, 4, 5, \dots,)$ are locally unstable.

Example 3.6 demonstrates the possibility that there are infinite many different cycles but only one of them is stable. To obtain a theoretical explanation, we need new concepts (Devaney, 1989, Chapter 1.11).

Let f be a function defined on I which has continuous third derivative. The *Schwarzian derivative* of function f is the following expression:

$$S(f) = \frac{f'''(x)}{f'(x)} - \frac{3f''(x)^2}{2f'(x)^2}.$$

If $f'(x) = 0$, then further analysis is needed to determine $S(f)$.

Point c is called a *critical point* of function f if $f'(c) = 0$. For example, for the logistic map, $S(f) = -6/(1 - 2x)^2$ and the unique critical point is $c = 1/2$. Note that $S(c) = -\infty$. Note also that if f and g are two smooth functions with negative Schwarzian derivatives, than the composite function $f \circ g$ has also a negative Schwarzian derivative. We can now formulate

Theorem 3.8. *(Singer, 1978.) If f is a smooth map of I into itself, it has a single critical point and its Schwarzian derivative is negative (including $-\infty$) in the entire interval I, then it has at most one stable cycle.*

Bifurcation

We shall generalize our observations on the stable fixed point and the limit cycles of periods 2 and 3 of the logistic function (Problem 3.3, Examples 3.5 and 3.6) and utilize Theorem 3.8.

Theorem 3.9. *Attractors of the logistic function. The logistic map* $x_t = ax_{t-1}(1 - x_{t-1})$ *has*

a) *a unique fixed point for* $1 < a < 3$,

b) *a unique* 2^k*-limit cycle and unstable* 2^h*-cycles* ($h = 1, 2, \ldots,$ $k - 1$) *for* $3 < a_k < a < a_{k+1} < a_\infty \approx 3.57$, *and*

c) *infinite number of cycles and uncountable number of aperiodic paths for* $a_\infty < a \leq 4$ *representing topological as well as truly chaotic dynamics.*

Remarks. 1. It occurs quite often that people forget about the uniqueness of limit cycles for any a with a logistic map. Uniqueness means that the other cycles are unstable, hence 'invisible'. Moreover, it may occur that the foregoing limit cycle is (almost) globally attracting: Example 3.6.

2. It is not clear a priori how the logistic map behaves in the chaotic interval (3.57,4]. Jakobson (1981) proved that the set of those parameters where the logistic map is truly chaotic has a positive measure.

Instead of proof. The proof is quite involved and requires strong mathematical tools, thus we only refer to the tent map of Theorem 3.6. The proof boils down to the following idea: increasing the value of parameter a, the relevant k-iterated function becomes similar to its predecessors, (Devaney, 1989).   ∎

A special tool of computer analysis is the so-called *bifurcation diagram* (Figure 3.6). It depicts *simultaneously* the *asymptotic* behavior of the system for many close parameter values, and highlights the qualitative changes. It can be described as follows.

Select an important parameter which can change in a relatively large interval. For example, let it be $a \in (2.7, 4)$ for the logistic equation. Let a lie on the horizontal axis and x_t ($t = U, U + 1, \ldots, V$) on the vertical axis. Let U and V be so large that the behavior for $t \in [U, V]$ characterizes well the long run behavior of the system, for example, $U = 100$ and $V = 200$. Let the division of the horizontal axis be sufficiently fine that an interesting value will not lie between two subsequent points: with step-size $h = 0.01$, $a(z) = a(z - 1) + h$.

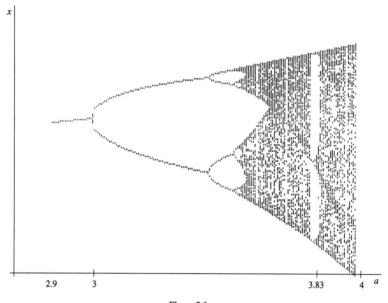

Figure 3.6
Bifurcation diagram: total

a) Let the initial value be $x_0 = 0.6$. b) The quality of the picture (and the dampening factor) can be greatly improved if the initial value of run a is equal to the end value of run $a - h$.

Problem 3.8. Bifurcation diagram. Write a computer program for the study of the asymptotic behavior of logistic equation.

Problem 3.9. Self-similarity. Magnify Figure 3.6 in the rectangle $3.841 < a < 3.857$ and $0.13 < x < 0.18$. Comparing Figures 3.6 and 3.7 yields that the small picture is similar to the large one. Repeating the magnification, similarity prevails: this phenomenon is called *self-similarity*.

If the foregoing phenomena were only characteristic to the logistic map, then not much discussion would be devoted to logistic map or bifurcation. However, these phenomena are universal and characterize all smooth unimodal functions.

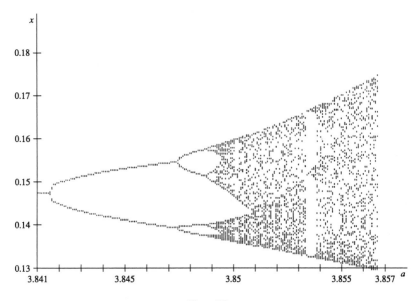

Figure 3.7
Bifurcation diagram: a detail

Ergodicity*

On the tent map one can illustrate that the individually chaotic paths
satisfy statistical laws. Among these laws the most important is the
ergodicity which generalizes the laws of large numbers (Day and Pia-
giani, 1991). Let ν be an absolutely continuous probability measure
(that is, a distribution having a measurable density function) on the
interval $[a, b]$ which is *f-invariant*:

$$\nu(f^{-1}(\mathcal{A})) = \nu(\mathcal{A}) \quad \text{holds for every measurable set} \quad \mathcal{A}.$$

Measure ν is called *ergodic* if the asymptotic time and space averages
of any integrable function g are equal:

$$\lim_{k} \frac{1}{k} \sum_{j=1}^{n} g(f^{j-1}(x)) = \int g \, d\nu, \quad \nu-\text{almost everywhere.}$$

For example, let $g = 1_{\mathcal{A}}$ be the *indicator function* of set $\mathcal{A}$, that is, $g(x) = 1$ for $x \in \mathcal{A}$ and 0 otherwise, and $v_k(x_0)$ be the relative frequency of the event $f^{j-1}(x) \in \mathcal{A}$. Then the relative frequency converges to the *theoretical value of the probability*: $\lim_k v_k(x_0) = \nu(\mathcal{A})$, ν-almost everywhere.

The following theorem provides a sufficient condition on the existence of an ergodic measure.

Theorem 3.10. *(See Grandmont, 1986, Theorem 5.)* *If there exists no stable cycle and there exists a neighborhood V of the critical point x^* such that a path starting from x^* never returns to the neighborhood, then there exists a unique absolutely continuous probability measure which is ergodic.*

Unfortunately, the ergodic measure can coexist with a quasi-cycle (Medio, 1995), thus ergodicity does not imply chaos.

Turbulence

A map $f : I \to I$ is called *turbulent* if interval I has two compact subintervals J and K, such that they have at most one joint point and their union is contained in intersection of their images: $J \cup K \subseteq f(J) \cap f(K)$. It is easily seen that $f(x) = 4x(1-x)$ as well as the tent map is turbulent for $J = [0, 1/2]$ and $K = [1/2, 1]$. It is noteworthy that such a simple notion implies such interesting properties.

Theorem 3.11. *a) If map $f : I \to I$ is turbulent, then f has a P-cycle for every integer $P > 1$.*

b) If map $f : I \to I$ has a P-cycle for at least one odd integer $P > 1$, then iterate f^2 is turbulent.

Analysis of complex dynamics in the plane

The analysis of complex dynamics is much more difficult than the stability analysis of equilibrium and cycle. Most of the economic papers on chaos simply rely on the well-known properties of the one-dimensional smooth maps. We shall also present such a model in Section 4.1. We go further and present models the simplest variants of which have two basic variables ($n = 2$), therefore their analyses have to apply more sophisticated tools (Hommes and Nusse, 1989 and Hommes, 1991).

A basic analytical tool is the *return map* which can loosely be defined as follows: Consider, for example, those t's where $x_{1,t} = x_1^u$ is a maximal value, and select two consecutive t's with that property: t' and t''. Denote J the range of $x_{2,t'}$ for all feasible initial states x_0 and then $x_{2,t''} = R[x_{2,t'}]$ is the return map. The behavior of f and R is related: both are chaotic (or cyclic) for the same parameter values (see Hommes, 1991).

4. DISCRETE-TIME NONLINEAR MODELS

The results of Chapters 1 and 2 made it evident that "life is too nice to be true" in linear deterministic dynamical systems. Relying on Chapter 3, we shall analyze time-invariant *discrete-time nonlinear economic models* in this Chapter. These models yield cycles and more complex behavior not only as exception but as rule. In Section 4.1 *irregular growth cycles* are analyzed by reduction to the logistic model of Section 3.3. In Sections 4.2–4.4 we shall generalize the linear models of Chapter 2 to nonlinear (piecewise linear) ones. (Please, go through them before turning to these Sections!) In Section 4.2 the *nonlinear accelerator–multiplier model* is discussed. In Section 4.3 we shall investigate the *nonlinear control of a multisector economy by stock signals*. Section 4.4* considers the macro variant of the expectation driven model. Section 4.5 introduces *mixed expectations*. Finally, Section 4.6 concludes.

4.1 IRREGULAR GROWTH CYCLES

This Section is devoted to a simple growth model (Day, 1982). Let y_t and k_t be the per capita *output* and *capital* at time t. Their connection is described by a neoclassical production function f:

$$(4.1) \qquad y_t = f(k_t).$$

Let c_t be the *consumption* and

$$(4.2) \qquad i_t = y_t - c_t$$

the *investment*. It will be assumed that investment is proportional to output:

$$(4.3) \qquad\qquad i_t = \iota y_t, \qquad \iota > 0.$$

Let d_t be the per capita *scrapping* of capital at time t. Then the dynamics of capital is given by

$$(4.4) \qquad\qquad k_t = k_{t-1} + \iota f(k_{t-1}) - d_t.$$

To simplify the analysis, it will be assumed that the scrapping is just equal to the capital inherited from the past:

$$(4.5) \qquad\qquad k_{t-1} = d_t.$$

Substitution of (4.5) to (4.4) yields the
General basic equation

$$(4.6) \qquad\qquad k_t = \iota f(k_{t-1}).$$

Day (1982) wanted to apply a logistic-like equation to growth cycles, thus he modified the well-known Cobb–Douglas production function:

$$(4.7) \qquad f(k) = Ak^\alpha \quad \text{where} \quad A > 0 \quad \text{and} \quad 0 < \alpha < 1.$$

He assumed that there exists a *maximal per capita capital* $k^* > 0$ where pollution is so high that zero production occurs. For lower level of capital, the positive effect of higher level of capital is weakened by a factor $(k^* - k)^\gamma$ where $\gamma > 0$. Thus the usual production function [(4.7)] is replaced by

$$(4.8) \qquad\qquad f(k) = Ak^\alpha (k^* - k)^\gamma.$$

Substitution of (4.8) into (4.6) yields the
Parametric basic equation

$$(4.9) \qquad\qquad k_t = \iota A k_{t-1}^\alpha (k^* - k_{t-1})^\gamma.$$

Obviously, the function at the R.H.S of (4.9) is unimodal. We shall assume that the parameter values ι, A, α, γ and k^* are such that (4.9) maps the interval $(0, k^*)$ into itself. Using the generalization of Theorem 3.9, we have

Theorem 4.1. *(Day, 1982.) Depending on the parameter values, stable, cyclic and chaotic paths emerge.*

Problem 4.1. Logistic map. Take $A = 20$, $\alpha = 1$, $k^* = 1$ and $\gamma = 1$. Show that for $\iota = 0.1$, 0.16 and 0.2, stable, 2-cyclic and chaotic paths emerge, respectively ($k_{-1} = 0.7$: Figure 4.1).

Problem 4.2. Transient chaos. Demonstrate on the computer that for $\iota = 0.1919$, a stable 3-cycles emerges, although for certain initial values, transient chaos prevails: ($k_{-1} = 0.538$, $k_{-1} = 0.539$: Figure 4.2).

Remarks.   1. The logistic map had been quite well-known before its connection with chaotic dynamics (Theorem 3.9) was discovered. For example, Samuelson (1947, p. 291) discussed it but in a continuous-time context.

2. We shall see in Example B.5 that Gale (1973) was quite close to discover chaos via a transformation of logistic maps in an Overlapping Generations model, but it was too early to uncover chaos.

3. Transforming a very popular continuous-time model of Goodwin (1967) into a discrete-time framework, Pohjola (1981) also applied the logistic equation to an interesting problem. His linear Phillips curve, however, yields negative wages for low employment. Pohjola only considered this fact as a numerical distortion, but in a necessarily global context, such a model is absurd. (It is an open question if there exists a nonlinear Phillips curve which generates *viable chaos*. I was only able to show that this is impossible via the linearly transformed logistic equation.)

4. Two surveys are mentioned here on the application of the logistic equations to dynamic economics: Cugno and Montrucchio (1984) and Boldrin and Woodford (1990). See also Section 8.3*.

4.2 NONLINEAR MODEL
OF ACCELERATOR–MULTIPLIER

In this Section we shall return to Hicks' trade cycle model, but now using the nonlinear version of the model discussed in Section 2.1. Modifying Samuelson (1939b), the basic idea of Hicks (1950) was that the planned linear control can be modified by exogenously given bounds, that is, the system is only piecewise linear, in other words: nonlinear. This method will be used in the next Sections, too.

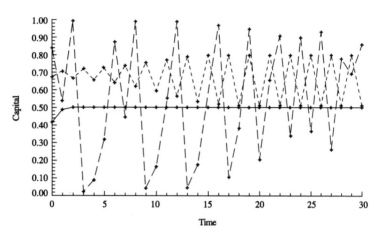

Figure 4.1
Fixed point, cycle and chaos

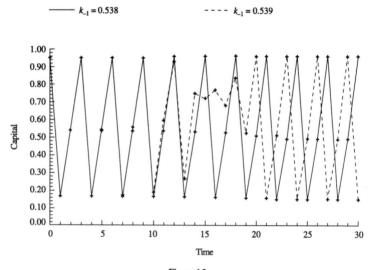

Figure 4.2
Transient chaos

SIMPLE MODEL

We immediately jump to the relative values: y_t, i_t and c_t are the detrended volumes of output, investment and consumption, respectively. Identity (2.1′) is simply copied, net investment and consumption equations (2.2′) and (2.3′) are taken as plans.

Output-identity

$$(4.10) \qquad y_t = i_t + c_t.$$

Planned investment function

$$(4.11) \qquad i_t^{\mathrm{P}} = i^{\mathrm{A}} + \beta\psi(y_{t-1} - \psi y_{t-2}).$$

Planned consumption function

$$(4.12) \qquad c_t^{\mathrm{P}} = c^{\mathrm{A}} + \psi\gamma y_{t-1}, \quad 0 < \gamma < 1.$$

As was mentioned earlier, Hicks introduced *floor i^{l}* of investment and *ceiling y^{u}* on output which constrain the corresponding plans.

Actual investment function

$$(4.13) \qquad i_t = \begin{cases} i^{\mathrm{l}} & \text{if } i_t^{\mathrm{P}} < i^{\mathrm{l}}; \\ i_t^{\mathrm{P}} & \text{otherwise.} \end{cases}$$

Actual consumption function

$$(4.14) \qquad c_t = \begin{cases} y^{\mathrm{u}} - i_t & \text{if } c_t^{\mathrm{P}} + i_t > y^{\mathrm{u}}; \\ c_t^{\mathrm{P}} & \text{otherwise.} \end{cases}$$

We have five equations and five variables. It is useful to eliminate the superfluous variables and equations. After some manipulation we obtain the

Basic equation

$$(4.15)$$
$$y_t = f_1(y_{t-1}, y_{t-2})$$
$$= \min\{\max[i^{\mathrm{A}} + \psi\beta(y_{t-1} - \psi y_{t-2}), i^{\mathrm{l}}] + c^{\mathrm{A}} + \psi\gamma y_{t-1}, y^{\mathrm{u}}\}$$

where y_{-2} and y_{-1} are given initial values.

In general, the equilibrium investment and output of the linear model should lie between the bounds of the nonlinear model.

Consistency

$$(4.16) \qquad i^{\mathrm{o}} > i^{\mathrm{l}} \quad \text{and} \quad y^{\mathrm{o}} < y^{\mathrm{u}}.$$

Local instability and limit cycle

The first statement of Hicks is presented without proof:

Theorem 4.2. *(Hicks, 1950.) If the equilibrium of the simple linear accelerator–multiplier model is stable, then its nonlinear counterpart is also stable.*

Example 3.4 demonstrates that local stability need *not* imply global stability. In the case of piecewise linearity, for one variable, local stability implies global stability, therefore one may think that this is generally true. We shall see in Section 4.6 that this is not the case: the introduction of control bounds may destroy stability.

Hicks presumed that actual systems are more or less cyclic, therefore he had to assume local instability for his model.

Claim. *(Hicks, 1950.) Assume that the nonlinear accelerator–multiplier model is locally unstable:*

$$(4.17) \qquad\qquad \psi^2 \beta > 1.$$

Then there exists a simple limit cycle, which is reached after the first hit of a bound.

Remarks. 1. It is easy to see that for degenerate oscillations, the system hits both the upper and the lower bounds. In case of regular oscillations, three cases are possible: the system hits a) both the upper and the lower bounds, b) only the upper bound and c) only the lower bound.

2. We shall call a cycle in the plane *simple* if the radius, connecting the state to the center, makes one turn while returns to its initial position. Otherwise we speak of a *composite cycle*.

To illustrate Hicks' model, we shall use the following data: $\psi = 1$ (zero growth), $y^u = 1.2$; $i^A = 0$, $c^A = 1 - \gamma$, that is, $y^o = 1$ (Blatt, 1983, p. 192).

Let us start with a simple limit cycle.

Example 4.1. Simple limit cycle. For $i^l = -0.1$; $\beta = 1.5$, $\gamma = 0.75$, there exists a simple limit 11-cycle (see Figure 4.3).

The following example provides a simple counter-example to Hicks' claim:

Example 4.2. Composite limit cycle. For $i^l = -0.1$; $\beta = 1.5$; $\gamma = 0.7$, there exists a composite limit 23-cycle which consists of two simple limit "11.5"-cycles (see Figure 4.4).

The following example provides a more complex but deeper counter-example to Hicks' claim:

Example 4.3. Quasi-limit cycle. For $i^1 = -0.05$; $\beta = 1.25$; $\gamma = 0.7$, there exists a quasi-limit cycle with (average) period "12.4" (see Figure 4.4).

After this prelude we announce

Theorem 4.3. *(Hommes, 1991, Theorem 4.1B.) Under the conditions of the Claim, the paths globally converge to one of the following attractors: simple, composite or quasi-cycle.*

Remarks. 1. The proof is quite involved (Hommes, 1991, Appendix to Chapter 4). In his proof Hicks probably overlooked that in a second-order system it is (y_t, y_{t-1}) rather than y_t which has to return. Fortunately, in economic terms there is not much difference between cycles and quasi-cycles, in both cases the average rotation number exists.

2. Hicks himself emphasized that in his model the upswing is faster than the downswing, just the opposite what happens in reality.

COMPOSITE MODEL

Already Hicks (1950) studied the composite model where the planned investment function or the planned consumption function is based on distributed lags. Due to the mathematical difficulties, he did not pay much attention to its analysis. It was Hommes (1991 and 1993) who made up this omission. First we copy (2.11) and (2.12), only adding superscript P.

Planned investment function with distributed lags

$$(4.11') \qquad i_t^{\mathrm{P}} = i^{\mathrm{A}} + \sum_{k>0} \psi^k \beta_k (y_{t-k} - \psi y_{t-k-1}).$$

Planned consumption function with distributed lags

$$(4.12') \qquad c_t^{\mathrm{P}} = c^{\mathrm{A}} + \sum_{k>0} \psi^k \gamma_k y_{t-k}, \quad \gamma_k \geq 0; \qquad \sum_{k>0} \psi^k \gamma_k < 1.$$

Hommes (1991, Chapter 4) proved that in the composite model chaotic paths also may occur. Here the following example suffices:

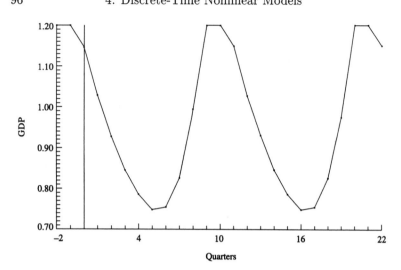

Figure 4.3
Hicksian 11-cycle

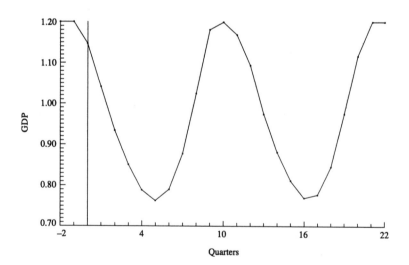

Figure 4.4
Hicksian double cycle

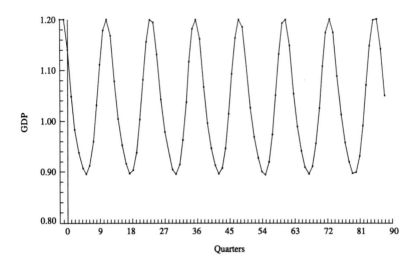

Figure 4.5
Hicksian quasi-cycle

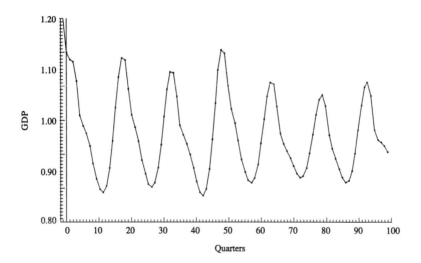

Figure 4.6
Hicksian chaos

Example 4.4. Chaos. For $y^u = 1.5$; $i^A = 0$, $\gamma_1 = 0.1$; $\gamma_2 = 0.3$; $\gamma_3 = 0.4$; $\gamma = \gamma_1 + \gamma_2 + \gamma_3 = 0.8$, $\gamma^A = 1 - \gamma$, $i^l = -0.1$; $\beta_1 = 2.25$; $\beta_2 = 0$, the composite Hicksian system behaves chaotically (see Figure 4.6).

Problem 4.3. Write a computer program and check graphically Examples 4.3–4.6.

Remarks. 1. It is the irony of fate that Hicks himself did not like his model just because he believed that, in contrast with reality, it behaves too regularly. Perhaps he would have been pleased to learn that his model does not behave nicely mathematically or does not behave too orderly economically.

2. Blatt (1978) and (1983, Chapter 11) made a very interesting trial. He set up a simple Hicksian model and estimated the generated time series by a linear econometric technique. The obtained linear stochastic equation hid the original nonlinear deterministic structure.

3. Hommes (1991, Chapter 4, p. 226) presented coexistence of limit cycles with periods 4 and 6. (Unfortunately, the orders of magnitudes are unrealistic.)

4.3 NONLINEAR MODEL OF CONTROL BY STOCK SIGNALS

In this Section we return to the model of control by output stock signals of Section 2.2, but to its nonlinear version. As was mentioned earlier, the basic idea of control by input- and output stocks is due to Kornai and Martos (1973) and its nonlinear version is discussed by Simonovits (1981a). The present discussion is confined to the control by output stock signals, thus it is new.

The variables and equations of the model

The nonlinear version differs from its linear predecessor in the nonlinearity of the output rule. For example, the linear output rule is only considered as a plan, and the actual output is constrained by lower and upper bounds. We just list the variables: y_t = output vector, z_t = output stock vector, both in time-period t, c = consumption vector and A = input-output matrix. We can now present the equations of the simple model.

Change in stocks

$$(4.18) \qquad\qquad z_t = z_{t-1} + (I - A)y_t - c.$$

Decentralized control of output

$$(4.19) \qquad y_{i,t} = y_i(z_{i,t-1}), \qquad i = 1, \dots, n$$

where $y_i(\cdot)$ is a nonincreasing function.

Two realizations are shown.

a) Fixing a stationary point (y°, z°), the nonlinear control rule may be

$$(4.20) \qquad y_{i,t} - y_i^\circ = \arctan[d_i(z_{i,t-1} - z_i^\circ)], \qquad i = 1, \dots, n.$$

b) Using the Hicksian idea:

Planned output

$$(4.21) \qquad y_t^P = y^* - \langle d \rangle z_{t-1}$$

where $\langle d \rangle > 0$ is a diagonal matrix and y^* is the capacity vector.

Actual output

$$(4.22) \qquad y_{i,t} = \begin{cases} y_i^l & \text{if } y_{i,t}^P \le y_i^l; \\ y_{i,t}^P & \text{if } y_i^l < y_{i,t}^P < y_i^u, \\ y_i^u & \text{if } y_{i,t}^P \ge y_i^u; \end{cases} \qquad i = 1, \dots, n$$

where y_i^l and y_i^u are the lower and upper bounds on output y_i, respectively.

Substituting (4.22) into (4.21), yields $y(z_{t-1})$. Substituting (4.19) into (4.18), provides the

Basic equation

$$(4.23) \qquad z_t = z_{t-1} + (I - A)y(z_{t-1}) - c.$$

Normal state and control

Let us start with the discussion of fixed point. For any possible y° and function $y(\cdot)$, $y^\circ = (I - A)^{-1}c$. Substituting into (4.19) and taking the inverse function of y_i, denoted by y_i^{-1}:

$$(4.24) \qquad z_i^\circ = y_i^{-1}(y_i^\circ) > 0, \qquad i = 1, \dots, n.$$

We have now

Theorem 4.4. *For (4.24), the nonlinear model controlled by output stock signals has unique normal state and control vectors which are positive.*

Global stability

Due to the simple structure of the model, under our assumptions the results of Section 3.1 directly apply. We shall need the assumption generalizing (2.24), that is, the control is a contraction:

$$(4.25) \qquad |y_i(z_i') - y_i(z_i)| \leq \delta|z_i' - z_i|, \quad 0 < \delta < 1.$$

Theorem 3.4 and Remark 3 belonging to it imply

Theorem 4.5. *Under assumption (4.25), the nonlinear control by stock signals is globally stable.*

Proof. Let us turn to the deviation vectors:

$$(4.26) \qquad z_t^{\mathrm{d}} = z_t - z_t^{\mathrm{o}} \quad \text{and} \quad y_t^{\mathrm{d}} = y_t - y_t^{\mathrm{o}}.$$

Then (4.23) is transformed into

$$(4.27) \qquad z_t^{\mathrm{d}} = z_{t-1}^{\mathrm{d}} + (I - A)y^{\mathrm{d}}(z_{t-1}^{\mathrm{d}})$$

where $y^{\mathrm{d}}(z_{t-1}^{\mathrm{d}}) = y(z_{t-1}) - y^{\mathrm{o}}$. It is obvious that y^{d} is also a contraction with a constant δ and it assigns an output above (below) the norm to an output stock below (above) the norm. Choose units of measurement such that all the column sums of A are less than unity (Theorem A.9):

$$(4.28) \qquad \mathbf{1}^{\mathrm{T}}A \leq \alpha\mathbf{1}^{\mathrm{T}}, \quad 0 < \alpha < 1.$$

Consider the coordinate-form of (4.27), take the moduli and add them up.

$$(4.29) \qquad \begin{aligned} \sum_i |z_{i,t}^{\mathrm{d}}| &\leq \sum_i \left[|z_{i,t-1}^{\mathrm{d}}| - (1 - \alpha)\delta|z_{i,t-1}^{\mathrm{d}}|\right] \\ &= [1 - (1 - \alpha)\delta] \sum_i |z_{i,t-1}^{\mathrm{d}}|. \end{aligned}$$

Using the l_1-norm in (A.15), (4.29) yields

$$(4.30) \qquad ||z_t^{\mathrm{d}}||_1 \leq [1 - (1 - \alpha)\delta] ||z_{t-1}^{\mathrm{d}}||_1,$$

hence Theorem 3.4 applies.   ∎

Problem 4.4. Stabilization of time-variant systems. Assume that the input-output matrix A varies in time: A_t, but there exists a constant system of measurement units for which the analogy of (4.28) holds:

$$(4.28') \qquad \mathbf{1}^{\mathbf{T}} A_t \leq \alpha \mathbf{1}^{\mathbf{T}}, \quad 0 < \alpha < 1.$$

Prove the generalization of Theorem 4.5.

Problem 4.5.* Work out the nonlinear variant of the model controlled by output and input stock signals.

4.4* NONLINEAR STOCK CONTROL WITH EXPECTATIONS

It would be nice to generalize the decentralized linear control model with expectation of Section 2.3* to a nonlinear one as was possible with the linear stock-signalling model of Section 2.2 in Section 4.3. Unfortunately, we are unable to do this, but in a macromodel we can handle the nonlinearity. The model was analyzed by Honkapohja and Ito 1980), Simonovits (1982), Hommes and Nusse (1989), Hommes (1991, Chapters 2A and 2B).

Disequilibrium dynamics

In disequilibrium theory, supply and demand are always distinguished, notations: superscript [S] and [D], respectively. Further, actual transaction is equated to the minimum of supply and demand. Consider one good market with nonperishable good and one labor market and assume that one unit of labor produces one unit of good. Moreover, introduce *sales expectation* formed at $t - 1$ for t and incorporate it to the behavior rules like in Section 2.3*. As a result, we have the following equations.
Supply of labor

$$L_t^{\mathrm{S}} = 1.$$

Supply of output
$$Y_t^S = I_{t-1} + L_t.$$

Inventory dynamics
$$I_t = I_{t-1} + L_t - Y_t.$$

Planned inventory
$$I_t^P = I_{t-1} + L_t^D - {}_{t-1}Y_t^D.$$

Demand for output
$$Y_t^D = \alpha + \mu L_t, \qquad \alpha > 0, \quad 0 < \mu < 1.$$

Normal inventory
$$I_t^* = \beta\,{}_{t-1}Y_t^D, \qquad \beta > 0.$$

Rational expectations for demand
$${}_{t-1}Y_t^D = Y_t^D.$$

Naive expectations for demand
$${}_{t-1}Y_t^D = Y_{t-1}^D.$$

Labor
$$L_t = \min(L_t^D, L_t^S).$$

Output
$$Y_t = \min(Y_t^D, Y_t^S).$$

For simplicity, we shall assume that planned and normal inventories are equal: $I_t^P = I_t^*$.

Rational expectations

Honkapohja and Ito (1980) and Hommes (1991, Chapter 2B) considered rational expectations (in deterministic framework, we could speak of perfect foresight).

We shall need the following notations:

$$\gamma = 1 - \mu - \alpha, \quad \chi = 1 - (\beta + 1)\mu, \quad I^u = (\beta + 1)\alpha, \quad I^l = I^u - \chi.$$

To avoid economically uninteresting cases, we shall assume that $\gamma > 0$ and $\chi > 0$.

After some calculations we obtain the following dynamics:

Labor

$$(4.31) \qquad L_t = \begin{cases} 1 & \text{if } I_{t-1} \leq I^l; \\ (I^u - I_{t-1})/\chi & \text{if } I^l < I_{t-1} < I^u; \\ 0 & \text{if } I_{t-1} \geq I^u. \end{cases}$$

With substitution, we obtain the following one-variable, piecewise linear difference equation:

Inventory dynamics

$$(4.32) \qquad I_t = \begin{cases} I_{t-1} + \gamma & \text{if } I_{t-1} \leq I^l; \\ \beta(\alpha - \mu I_{t-1})/\chi & \text{if } I^l < I_{t-1} < I^u; \\ I_{t-1} - \alpha & \text{if } I_{t-1} \geq I^u. \end{cases}$$

Steady state

A simple computation yields the *Keynesian* steady state:

$$I^\circ = \frac{a\beta}{1 - \mu} \quad \text{and} \quad L^\circ = \frac{\alpha}{1 - \mu}.$$

It is obvious that the steady state is independent of the type of expectations, since at the steady state every reasonable expectation scheme provides the same expectation, namely, the equilibrium.

Stability and instability

First we shall investigate stability.

Theorem 4.6. *(Honkapohja and Ito, 1982.) Under rational expectations, the steady state of the macro control (4.32) is locally stable if and only if $\mu\beta < \chi$.*

Remark. Remember that for every one-variable piecewise linear system, local stability implies global stability as well.

Proof. Evident from (4.32). ∎

What happens if the system is unstable?

Theorem 4.7. *(Hommes, Theorem 2B.1.) Under rational expectations, the macro control (4.32) is chaotic if $\mu\beta > \chi$.*

The proof is complex, it is omitted.

Naive expectations

Simonovits (1982), Hommes and Nusse (1989) and Hommes (1991, Chapter 2A) studied the naive expectations.

We shall need the following modified notations:

$$I^u(L_t) = (\beta + 1)(\alpha + \mu L_t), \quad I^l(L_t) = I^u(L_t) - \chi.$$

After some manipulations, the separation of cases yields the following, piecewise linear, planar difference equation:

Labor dynamics

$$(4.33) \qquad L_t = \begin{cases} 1 & \text{if } I_{t-1} \leq I^l(L_{t-1}); \\ L_t^P & \text{if } I^l(L_{t-1}) < I_{t-1} < I^u(L_{t-1}); \\ 0 & \text{if } I_{t-1} \geq I^u(L_{t-1}), \end{cases}$$

Inventory dynamics

$$(4.34) \qquad I_t = \begin{cases} I_{t-1} + \gamma & \text{if } I_{t-1} \leq I^l(L_{t-1}); \\ I_t^P & \text{if } I^l(L_{t-1}) < I_{t-1} < I^u(L_{t-1}); \\ I_{t-1} - \alpha & \text{if } I_{t-1} \geq I^u(L_{t-1}) \end{cases}$$

where

$$L_t^P = \frac{I^u(L_{t-1}) - I_{t-1}}{\chi}, \quad I_t^P = \mu(\beta+1)[(1-\mu)L_{t-1} - \alpha] + \mu I_{t-1} + \alpha\beta.$$

Theorem 4.8. *Under naive expectations, the steady state of (4.33)–(4.34) is stable if and only if $\chi > 0$.*

Sketch of the proof. a) Local stability: Using Theorem 1.11 it can be proved that the system regularly oscillates and is stable if $\mu(\beta + 1) < 1$. b) Global stability: With geometrical reasoning it can be shown that local stability implies global stability. The basic idea is as follows: extend the path from discrete points to continuous curves and observe that the paths, developing after hitting the wall of full employment, jump to paths closer to the equilibrium. ∎

What happens if the system is unstable? In Simonovits (1982) I risked the conjecture that locally unstable systems are almost always chaotic. Hommes and Nusse (1989) disproved my conjecture.

Let us see two nice examples where $\alpha = 0.95$ and $\mu = 0.9$.

Example 4.5. Limit cycle. $\beta = 0.3$; $L_{-1} = 0.96$; $I_{-1} = 0.28$.

Example 4.6. Acyclic chaos. $\beta = 0.4335$: $L_{-1} = 1$; $I_{-1} = 0.3$.

We present without proof

Theorem 4.9. *(Hommes, 1991, Theorem 2A.1.) Under naive expectations, the locally unstable system may have a limit cycle, quasi-cycle and chaos.*

Comparing the stability conditions of the two expectations, it is immediately obvious that the stability of rational expectations implies that of naive expectations. Note that this result qualitatively contradicts both our intuition and Theorem 2.7d. Two conclusions can be drawn. a) The perfectness of rational expectations is not absolute, even they can be worse than the much debased predecessors, the naive expectations. b) Our results are very sensitive to the apparently unimportant elements of our models, namely, if there are input stocks or not.

4.5. MIXED EXPECTATIONS

In this Section we shall consider an abstract dynamical system (Grandmont and Laroque, 1990 and Grandmont, 1998) driven by a combination of rational and naive, that is, mixed expectations (d-expectations, Molnár and Simonovits, 1998).

Let time be denoted by $t = 0, 1, 2, \ldots$ and the state of the system at t by a scalar x_t. Expectations with various horizons will play a crucial role in the analysis. $_t x_\tau$ will denote the expected state at time-period t for future time-period $\tau (> t)$. Let m and n be two positive integers, representing the number of past states and future expectations, respectively, influencing the dynamics.

The dynamics of the model is as follows:

$$(4.35) \qquad g\big(x_{t-m}, \ldots, x_{t-1}, \; x_t, \; _t x_{t+1}, \ldots, \; _t x_{t+n}\big) = 0.$$

Mixed expectations

Before defining the central concept of this Section, namely, mixed expectations, we shall recapitulate two well-known special cases, rational expectations and naive expectations.

Rational expectations

Each future expected state is equal to the *actual* (future) state:

$$(4.36) \qquad _t x_{t+i} = x_{t+i}; \qquad i = 1, \ldots, n.$$

Naive expectations

Each future expected state is equal to the *current* state:

$$(4.37) \qquad _t x_{t+i} = x_t; \qquad i = 1, \ldots, n.$$

Mixed expectations

We make the following generalization of these two types of expectations. Let d be an integer: $0 \leq d \leq n$.

(i) In addition to the current state x_t, the *near* future states $x_{t+1}, \ldots, x_{t+d-1}$ are known at date t:

$$(4.38) \qquad _t x_{t+i} = x_{t+i}, \qquad i = 1, \ldots, d.$$

(ii) Expectations on *far* future states $x_{t+d+1}, \ldots, x_{t+n}$ are equal to the state at date $t + d$:

$$(4.39) \qquad _t x_{t+i} = x_{t+d}, \qquad i = d + 1, \ldots, n.$$

Remark.   Mixed expectations are mixed from rational and naive expectations. (4.38)–(4.39) is indeed a generalization, since rational and naive expectations are obtained for $d = n$ and 0, respectively. Note also that for $n = 1$, (i)–(ii) collapse to either rational expectations ($d = 1$) or naive expectations ($d = 0$). We shall call d the *degree of rationality*.

Inserting (4.38)–(4.39) into (4.35) yields the *basic equation*:

$$(4.40) \qquad g\big(x_{t-m}, \ldots, x_{t+d-1}, x_{t+d}, \ldots, x_{t+d}\big) = 0.$$

It is a remarkable feature of our mixed expectations that at time-period t, x_{t+d} rather than x_t is determined. Apart from naive expectations, at time-period 0, not only past states $x_{-m}, \ldots, x_{-1}$ but also the initial state x_0 and the near future expectations $_0 x_1, \ldots, {}_0 x_{d-1}$ have to be given to obtain a path. Laitner (1981) calls the past states *historical initial values* and the present state and the future expectations *nonhistorical initial values*. The higher the degree of rationality, the more problematic becomes the arising wedge.

We shall introduce the notion of steady state and then turn to the local analysis around a steady state.

Local stability

Again, denote by x° a steady state. By definition,

$$g(x^\circ, \ldots, x^\circ) = 0.$$

It is easy to see that any steady state is independent of the value of d.

From now on we shall assume the existence of at least one steady state. (Note that in Appendices B and C there are typically two or more steady states.)

We linearize (4.40) around x°. Let γ_i be the partial derivative of g with respect to x_{t+i}, $i = -m, \ldots, n$ at point x°. Let $x_t^{\mathrm{d}} = x_t - x^\circ$. Then the local g_d-dynamics around x° is described by

$$\sum_{i=-m}^{d-1} \gamma_i x_{t+i}^{\mathrm{d}} + \left(\sum_{j=d}^{n} \gamma_j \right) x_{t+d}^{\mathrm{d}} = 0.$$

To get rid of negative indices, introduce $\alpha_i = \gamma_{i-m}$. The characteristic polynomial of the local dynamics g_d is

$$p_d(\lambda) = \sum_{i=0}^{m+d-1} \alpha_i \lambda^i + \left(\sum_{j=d}^{n} \alpha_{m+j} \right) \lambda^{m+d}.$$

Stability only depends on the location of the characteristic roots: Theorem 3.3.

The dynamics may be very complex, even around a steady state it is determined by the roots of an $(m+d)$-degree polynomial and $m+d$ initial conditions.

Obviously, there is one degree of freedom in selecting function g or correspondingly, the α_is. We shall normalize the latter as $\alpha_{m+n} = 1$.

The following example illustrates the situation in the simplest case.

Example 4.7. (See Grandmont, 1998.) Let $m = n = 1$. Normalization: $\alpha_2 = 1$. Notation: $\beta = \alpha_1 + 1 \neq 0$. Then $p_0(\lambda) = \alpha_0 + \beta\lambda$ and $p_1(\lambda) = \alpha_0 + \alpha_1\lambda + \lambda^2$. The stability conditions of naive expectations are trivial: $-1 < \lambda = -\alpha_0/\beta < 1$, that is, stability is equivalent to $|\alpha_0| < |\alpha_1 + 1|$. The stability conditions of rational expectations are a little bit more complex (see Example 1.12): $p_1(1) > 0$, $p_1(-1) > 0$

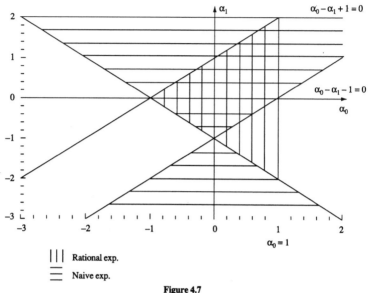

Figure 4.7
Stability conditions for expectations

and $p_1(0) < 1$, that is, $\alpha_0 + \alpha_1 + 1 > 0$, $\alpha_0 - \alpha_1 + 1 > 0$ and $\alpha_0 < 1$. Figure 4.7 illustrates the situation in the (α_0, α_1)-plane. Areas shaded vertically and horizontally represent stability of rational and naive expectations in the parameter plane, respectively. Their joint set, by default, depicts simultaneous stability.

Turning to the conditions of saddle-point instability: either $p_1(1) < 0 < p_1(-1)$ or $p_1(-1) < 0 < p_1(1)$, that is, $|\alpha_1| > |\alpha_0 + 1|$. Denoting by λ_2 the stable root and choosing $x_0^{\mathrm{d}} = \lambda_2 x_{-1}^{\mathrm{d}}$, eliminates the explosive direction. (Note, however, that any conceivable remaining error will eventually destroy stability unless one works in backward direction which is quite unnatural.)

What happens, however, if both roots are unstable? Then even the rather shaky trick is weakened and there is no way out of instability under rational expectations except constraining the system to start from the steady state. But the economically unjustified distinction between the two types of instability undermines the whole procedure. Therefore we generally consider stability without constraint.

For the time being, we have only a rather simple set of sufficient conditions on the stability/instability of mixed expectations.

Theorem 4.10. *For given d, assume that mixed expectations have no unit roots.*

a) *The steady state is unstable if*

$$(4.41) \qquad |\alpha_0| \geq \left| \sum_{j=d}^{n} \alpha_{m+j} \right|$$

b) *and it is stable if*

$$(4.42) \qquad \sum_{i=0}^{m+d-1} |\alpha_i| \leq \left| \sum_{j=d}^{n} \alpha_{m+j} \right|.$$

Remarks. 1. The no unit root assumption is typical in the literature and it is generic.

2. Assumption (4.41) means that in absolute terms the first effect of the past is stronger than or equal to the last $n - d + 1$ effects of the future.

3. Assumption (4.42) is quite difficult to interpret in general. If the sign of α_is is alternating, then (4.42) can hardly hold.

Proof. a) The instability condition of mixed expectations is almost trivial. Our polynomial p_d is of $(m + d)$-degree. Recall the well-known formula that the modulus of the product of the roots of p_d is equal to the ratio of the R.H.S to the L.H.S. of (4.41). Hence the modulus is equal to or larger than 1. By assumption, no root lies on the unit circle line, that is, at least one of them lies outside the unit circle.

b) The stability condition of expectations is also elementary. Assume the opposite, that is, there exists an unstable root of p_d, say λ_1 for which $|\lambda_1| > 1$. Then introducing $\beta_d = \sum_{j=d}^{n} \alpha_{m+j}$,

$$p_d(\lambda_1) = \sum_{i=0}^{m+d-1} \alpha_i \lambda_1^i + \beta_d \lambda_1^{m+d}, \quad \text{that is,} \quad -\beta_d \lambda_1^{m+d} = \sum_{i=0}^{m+d-1} \alpha_i \lambda_1^i.$$

Taking the absolute values, dividing by $|\lambda_1|^{m+d}$ and applying the triangle inequality, yields

$$|\beta_d| \leq \sum_{i=0}^{m+d-1} |\alpha_i| \, |\lambda_1|^{i-m-d}.$$

Because $|\lambda_1|^{i-m-d} < 1$, dropping them only increases the R.H.S.:

$$|\beta_d| < \sum_{i=0}^{m+d-1} |\alpha_i|,$$

contradicting (4.42).   ∎

If $d = n$ and $d = 0$ are substituted into the instability condition (4.41) and the stability condition (4.42), then we obtain the sufficient conditions of the simultaneous instability of rational expectations and stability of naive expectations, respectively.

Theorem 4.11. *Assume that neither $p_n(\lambda)$ nor $p_0(\lambda)$ has unit roots.*

a) Under rational expectations, the steady state is saddle-point unstable if

$$(4.43) \qquad\qquad |\alpha_0| \geq \alpha_{m+n} = 1.$$

b) Under naive expectations, the steady state is stable if

$$(4.44) \qquad\qquad \sum_{i=0}^{m-1} |\alpha_i| \leq \left| \sum_{j=0}^{n} \alpha_{m+j} \right|.$$

Remarks.   1. Assumption (4.43) means that in absolute terms the first effect of the past is stronger than or equal to the last effect of the future.

2. To interpret assumption (4.44), let us assume that all the past effects have the same sign: either $\alpha_i \geq 0$, $i = 0, \ldots, m - 1$ or $\alpha_i \leq 0$, $i = 0, \ldots, m - 1$. Then by introducing

$$\alpha = \sum_{i=0}^{m-1} \alpha_i \quad \text{and} \quad \beta = \sum_{j=0}^{n} \alpha_{m+j},$$

(4.44) simplifies into

$$(4.45) \qquad\qquad |\alpha| \leq |\beta|.$$

Note that we have already met (4.45) without equality in Example 4.7 (see also Grandmont (1998, Proposition 2.2)).

It is remarkable that a good deal of authors are pleased rather than embarrassed with the instability arising with rational expectations. The often arising *balanced saddle-point instability* (when the number of unstable directions and that of undetermined expectations are the same) ensures that the nonhistorical initial values are determined so as to eliminate the unstable directions: *determinacy* (Laitner, 1981). The simplest way to obtain determinacy is to assume that the effects of the past and the future are symmetric:

$$(4.46) \qquad m = n \quad \text{and} \quad \alpha_{2n-i} = \alpha_i, \qquad i = 0, \dots n - 1.$$

This symmetry is easily obtained if function g is already symmetric, that is, the dynamics is reversible in time. (It is recalled that reversibility occurs in mechanics but not in thermodynamics.)

For that reason the following proposition is a useful complement to Theorem 4.11.

Theorem 4.12. a) *Under rational expectations, the steady state is a balanced saddle-point, hence determinate if the symmetry conditions (4.46) are satisfied.*

b) *With symmetry and nonnegativity, the dynamics under naive expectations is stable if either $\alpha_n > 0$ or $\alpha_n < 0$ and $\alpha < -\alpha_n/2$.*

Proof.   a) The balancedness of the saddle-point of rational expectations is almost trivial. By (4.46), our polynomial $p_n(\lambda)$ is *reciprocal*, that is, if λ is a root, then $1/\lambda$ is also a root. Thus the number of unstable directions is n, and so does the number of nonhistorical initial conditions.

b) Now $\beta = \alpha_n + \alpha$, and (4.45)–(4.46) imply one of the conditions. ∎

Does the partial failure of rational expectations make the assumption of naive expectations acceptable? Is it not true that naive expectations mean persistent trivial error? In the classical case of cobweb cycle this is indeed the case, but there are other, nonlinear dynamical systems where no linear statistical method discovers any systematic error (Hommes and Sorger, 1997).

4.6 RELATED ISSUES

Before closing this Chapter on nonlinear deterministic systems, we return to some problems already touched in the Introduction and

Sections of the present Chapter. To get rid of problems of linear deterministic models, the researchers of business cycles have chosen either of the two routes: (i) stochastic linear models (Frisch, 1933), or (ii) deterministic nonlinear models (Hicks, 1950). The mainstream economists have sided with the first approach, but some influential economists adhered to the second one. Without modern nonlinear mathematical methods and good computer graphics, however, their results remained rather limited. It was only around 1980 that deterministic nonlinear dynamical models started to exert a significant influence.

Both approaches have their strong and weak points. Cycles in the economic life are much less regular than in nature. Stochastic models are flexible enough to reflect economic irregularities. (Remember Hicks' disappointment with his exact cycles.) Chaos theory, however, made deterministic nonlinear models capable to describe erratic dynamics as well.

In real life economies the upswing lasts much longer than the downswing. As Blatt (1980) and (1983, Chapter 10) pointed out, stochastic linear models show symmetry – at least on the average, contradicting real asymmetry. Of course, nonlinear deterministic models are capable to describe asymmetric oscillations. We simply refer to Brock (1986) for further thoughts on this comparison. The difference between the two approaches has an interesting policy implication. Roughly speaking, the adherents of the stochastic approach oppose government intervention into the economy, those of the nonlinear approach support it. This situation is quite logical. The effects of stochastic disturbances cannot be mitigated, because of significant lags and measurement errors. This is not the case with nonlinear deterministic jumps. In this Chapter we have followed the nonlinear approach, but we shall turn to the stochastic one in Chapters 7 and 8.

At the end I venture some remarks on the relation among the piecewise linear models just presented. Samuelson (1947) called them *billiard table* models. The starting point is Hicks' theorem on the preservation of stability and his claim on convergence to limit cycles. In Simonovits (1981a) I discovered the possibility that control systems with bounds may lose their stability (destabilization). Considering a nonlinear stock-signalling model of Honkapohja and Ito (1980), in Simonovits (1982) (see above Section 4.4) I risked the following conjecture: Every locally unstable piecewise linear model is almost surely chaotic. This conjecture was as wrong as Hicks' claim, representing two extreme positions. Hommes and Nusse (1989), Hommes (1991, Chapter 2A) and Hommes (1992) provided the balanced analysis of Honkapohja and Ito (1980), while Hommes (1991, Chapter 4) and (1992) did the same with Hicks' model.

Having the insight of Hommes and Nusse (1989), Hommes (1991, Chapter 3), Simonovits (1991), (1992, Appendix 2), and Hommes et al. (1995) came up with a nonlinear model of socialist investment cycles as formulated originally by Bauer (1978), Lackó (1980) and Kornai (1982). As a by-product, Hommes and Nusse (1992) analyzed the *border-crossing bifurcation*.

My simile of billiard table was wrong anyway: apart from exceptional cases, a billiard ball on a billiard table moves along a quasi-cycle.

At the end, it may be useful to remind ourselves of the difference between the mathematical and economic concepts of cycle. If we accept Ickes' (1986) economic concept (the triple of recurrence, reinforcement and regularity), then not only the limit cycles but also other, nondamped oscillations should be accepted as cycles.

5. ORDINARY DIFFERENTIAL EQUATIONS

We shall speak of *ordinary differential equation* if an equation contains time-dependent (or more generally, scalar-dependent) variables as well as their derivatives with respect to time (or another scalar). Since we shall always consider ordinary differential equations in this book, we shall drop the adjective *ordinary*. We shall see the deep similarity between the theorems on differential equations and those on difference equations. To avoid lengthy references to Chapters 1 and 3, we shall only 'repeat' the most important definitions and theorems with the necessary changes. But we will refrain from displaying all the parallels. In Section 5.1 we shall introduce the *basic concepts* of differential equations. In Section 5.2 we outline the theory concerning *linear systems*. In Section 5.3 we shall return to *nonlinear systems* and extend stability theorems obtained for linear systems to nonlinear ones. In Section 5.4 we just outline the basic elements of *control* in continuous time. Useful information is to be found in Samuelson (1947), Coddington and Levinson (1955), Pontryagin (1962), Gandolfo (1971), Arnold (1973) and Martos (1981).

5.1 BASIC CONCEPTS AND THEOREMS

First-order vector differential equations

Let $\mathbf{T} = [0, T]$ be a (time) *interval* of the reals, x an n-vector and $f(\cdot, \cdot)$ be a $\mathbf{T} \times \mathbf{R}^n \to \mathbf{R}^n$ function. Let $x(t)$ be a smooth $\mathbf{T} \to \mathbf{R}^n$ function and $\dot{x}(t)$ be its derivative function. If x and $\dot{x}$ satisfy the

First-order vector differential equation

$$(5.1) \qquad \dot{x}(t) = \frac{dx}{dt} = f[t, x(t)]$$

in the interval $\mathbf{T}$, then x is called a *solution* of this system of equations. The initial state $x(0) = x_0$ is generally given and then we have a *problem with initial values*. But in some applications (see for example, Chapter 9), the end values or both initial and end values are given. Without such conditions, the problem would be indeterminate.

Form (5.1) is coordinate-free, which is very advantageous by its conciseness. It is quite common, however, that the system of equations is given in coordinates:

$$(5.1') \qquad \dot{x}_i(t) = f_i[t, x_1(t), \dots, x_n(t)], \qquad i = 1, \dots, n$$

where functions $f_i : \mathbf{R}^{1+n} \to \mathbf{R}$ and functions $x_i(t) : \mathbf{R} \to \mathbf{R}$.

Higher-order differential equations can be reduced to first-order equations the same way as difference equations can. Take $y^{(n)} = g(t, y^{(n-1)}, \dots, \dot{y}, y)$. Using notation $x_i = y^{(i-1)}$, $i = 1, \dots, n$, we have obtained the n-variable first-order system

$$\dot{x}_1 = x_2, \dots, \dot{x}_{n-1} = x_n, \qquad \dot{x}_n = g(t, x_{n-1}, \dots, x_2, x_1).$$

In contrast to difference equations, certain differential equations have no solutions, others have multiple solutions. It is also possible that the solution exists and is unique, but it cannot be extended to the entire time interval $\mathbf{T} = [0, \infty)$. (We shall mostly consider *explicit* differential equations. With implicit equations even difference equations may have no solution or multiple solution, see Section C.3.)

We present several examples.

Example 5.1. Nice solution. $\dot{x} = \lambda x$ with $x(0) = x_0 \Rightarrow x(t) = x_0 e^{\lambda t}$, $t > 0$.

Example 5.2. No solution. $\dot{x} = 1$ if x is rational, $\dot{x} = 0$ otherwise. (According to Darboux's theorem, a derivative function – even if it is not continuous – takes on any intermediate value, Rudin, 1976, Theorem 5.12.)

Example 5.3. Multiple solution. $\dot{x} = x^{2/3}$, $x_0 = 0 \Rightarrow x(t) = 0$ and $x(t) = (t/3)^3$ are two solutions.

Example 5.4. Bounded time interval for the solution. $\dot{x} = x^2$, $x(0) = x_0 \Rightarrow x(t) = -1/(t - 1/x_0)$ is not defined at $t = 1/x_0$.

If one knows a solution, then by substitution one can show that it is a solution. There are a large number of tricks to find a solution, but we shall hardly dwell on such techniques. Since many differential equations have no closed-form solution (with elementary functions), one has to resort to approximate numerical solutions.

Example 5.5. No closed-form solution (Arnold, 1973). The differential equation $\dot{x} = x^2 - t$ has no closed-form solution.

Example 5.6. Mathematical pendulum. Let g be the gravitational acceleration, L be the length of the pendulum and $y(t)$ be the current angle at time t. According to the laws of mechanics, $y(t)$ satisfies the following second-order nonlinear differential equation: $\ddot{y} = -B \sin y$, $B = g/L$, $y_0 = A$ (the amplitude of the pendulum), $\dot{y}_0 = 0$. This equation also lacks a closed-form solution. We know from experience and can prove mathematically that it has a periodic solution: $y(t) \equiv y(t - P_A)$ where P_A depends on A.

Existence and uniqueness

Having reviewed the examples and counter-examples, we present some positive theorems. Let $\mathcal{X} \subseteq \mathbf{R}^n$ and let the initial state $x_0 \in \mathcal{X}$, *feasible*.

First we give an existence theorem.

Theorem 5.1. *(Cauchy and Peano's theorem, 19th century.) If $f(t, x)$ is continuous on $\mathbf{T} \times \mathcal{X}$, then for any feasible initial condition, there exists at least one solution of (5.1) on a subinterval J of $\mathbf{T}$.*

Instead of proof.* The general proof is based on deep mathematical ideas. The basic idea of the proof goes back to Euler and is easy to explain and is based on the methods of difference equations (Chapters 1 and 3). First fix an integer k. Let us divide interval $\mathbf{T}$ to k equal subintervals with length $h = T/k$. Let us denote the equidistant division points by $t_{i,k}$, $i = 0, \ldots, k$. Because of continuity, $f(t, x)$ can be well approximated by $f(t_{i-1,k}, x_{i-1,k})$ on the interval $\mathbf{T}_{i,k} = (t_{i-1,k}, t_{i,k})$. Then one obtains an approximate solution from $x_k(t) = x_k(t_{i-1,k}) + f(t_{i-1,k}, x_{i-1,k})(t - t_{i-1,k})$. If $k \to \infty$, then a deep mathematical theorem (Rudin, 1976, Theorem 7.23) guarantees that a subsequence of $\{x_k(t)\}$ converges to a function $x(t)$, for every $t \in \mathbf{T}$. By construction, the limit function is a solution of the differential equation.   ∎

Using the foregoing method under $k = 20$, we have solved the differential equation of Example 5.5 for the initial value $x_0 = 1$ in the interval $[0;1]$ and displayed the approximate solution in Figure 5.1.

Problem 5.1. Broken lines. Consider the differential equation $\dot{x} = \lambda x$, $x_0 = 1$ of Example 5.1. Prove that for a fixed t, a) the k-part approximation by broken lines is $x_k(t) = (1 + \lambda t/k)^k$ and b) for $k \to \infty$, it converges to $e^{\lambda t}$. (It can be proved that the approximate solutions converge to the exact solution on the entire interval $[0, t]$.)

Using the foregoing method under $k = 20$ and 40, we have solved the differential equation of Example 5.1 for the initial value $x_0 = 1$ in the interval $[0;1]$ and displayed the exact and the approximate solutions in Figures 5.2 and 5.3 with parameter values $\lambda = -1$ and 1, respectively. As is expected, the finer the approximation, the better.

Second, we give a uniqueness theorem. We need, however, a stronger assumption than continuity. Function $f(t, x)$ is said to satisfy the *Lipschitz condition* on $\mathbf{T} \times \mathcal{X}$ where $\mathcal{X}$ is a convex domain of $\mathbf{R}^n$ if $f(t, \cdot)$ is continuous in t and there exists a positive real K for which $\|f(t, x) - f(t, y)\| < K\|x - y\|$ for all $t \in \mathbf{T}$ and $x, y \in \mathcal{X}$. It is well-known that if f is continuously differentiable on a compact domain $\mathbf{T} \times \mathcal{X}$, then f satisfies the Lipschitz condition. (Note that the Lipschitz condition was introduced in order to prove uniqueness of solutions of differential equations.)

Theorem 5.2. (*Picard and Lindelöf's theorem, 1890, 1894.*) If $f(t, \cdot)$ satisfies the Lipschitz condition on $\mathbf{T} \times \mathcal{X} \subseteq \mathbf{R}^{1+n}$, then for any feasible initial condition, the differential equation (5.1) has a unique solution in a subinterval J of $\mathbf{T}$.

Sketch of the proof.* The general proof is based on deep mathematical results (Coddington and Levinson, 1955, Chapter 1, Theorem 1.3). Nevertheless, the basic idea is easy to explain and worth discussing. First, using the Newton–Leibniz formula, we shall transform the differential equation (5.1) into an equivalent integral equation:

$$(5.2) \qquad x(t) = x(0) + \int_0^t f[\tau, x(\tau)] \, d\tau.$$

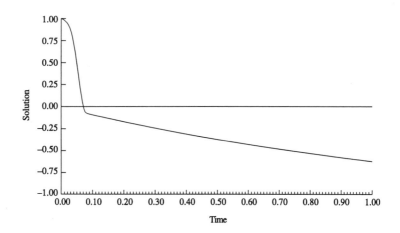

Figure 5.1
Approximation to $\dot{x} = x^2 - t$

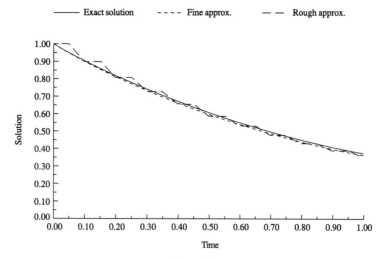

Figure 5.2
Approximate solutions of $\dot{x} = -x$

Let $x_k(t)$ be the kth approximate solution to the integral equation (5.2) where $x_0(t) \equiv x_0$. Then using the general iteration mentioned after Theorem 3.4, the next, $(k+1)$th approximate solution is given by

$$(5.3) \qquad x_{k+1}(t) = x(0) + \int_0^t f[\tau, x_k(\tau)]\, d\tau.$$

Here the idea of abstract space enters. Let the continuous function x be an element of the Banach space $C[J]$ of continuous functions where J is an appropriate subinterval of $\mathbf{T}$. Then the above iteration can be represented as $x_{k+1} = \phi(x_k)$ where ϕ is an abstract mapping from $C[J]$ to itself and $x_0(t) \equiv x_0$ (constant function). It can be demonstrated that ϕ is a contraction, that is, it has a unique stationary point, namely, the solution of the integral equation (5.2). ∎

In the theory of differential equations it is very important that *the solution continuously depends on the initial values*. This property can be deduced from the proof. In chaotic behavior (see below) this dependence becomes very weak, however.

Methods of solution

The method just described is called *successive approximation* which can be applied to any smooth function f. (At this point we have to give up our convention that the step variable of a discrete-time system is also denoted by t.)

Problem 5.2. Example for successive approximation. Consider the differential equation $\dot{x} = \lambda x$, $x_0 = 1$ of Example 5.1. Prove that the method of successive approximation gives a) the k-degree Taylor polynomial of $e^{\lambda t}$:

$$x_k(t) = \sum_{j=0}^{k} \frac{(\lambda t)^j}{j!}$$

b) which converges to the solution $e^{\lambda t}$ as $k \to \infty$.

Figure 5.4 displays the exact and the approximate solutions for $k = 1$, 2 and 4 ($\lambda = 1$). Observe that the error of approximation grows with time.

Problem 5.3. No Lipschitz condition. Show that the function appearing in Example 5.3 does not satisfy the Lipschitz condition.

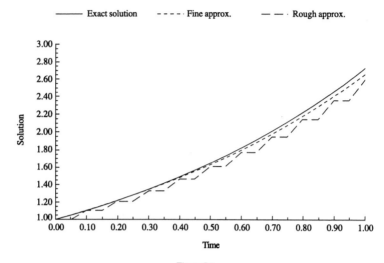

Figure 5.3
Approximate solutions of $\dot{x} = x$

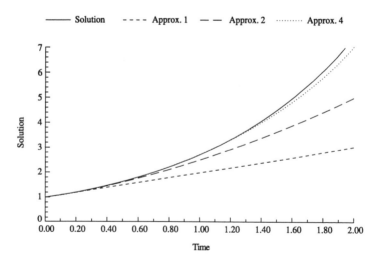

Figure 5.4
Successive approximations to $\dot{x} = x$

After presenting two general theorems on existence and uniqueness, we depict the solution of differential equations of the two simplest types.

Theorem 5.3. *Simple integrability. If $f(t, x)$ is independent of x, that is, $f(t, x) \equiv f(t)$, then the differential equation (5.1) is simply integrable:*

$$(5.4) \qquad x(t) = x(0) + \int_0^t f(\tau) \, d\tau.$$

Proof. Follows from the equivalent integral equation (5.2). ∎

Remarks. 1. Apparently, the theorem concerns systems. In fact, we have n independent scalar equations.

2. In this case, the successive approximation described in Theorem 5.2 gives the exact solution in the first step, thus there is no need for further steps.

Example 5.7. *Free fall.* As is known since Galilei, the speed of a mass point, freely falling in a homogeneous gravitational space, is $\dot{x} = \dot{x}_0 + gt$ where $\dot{x}_0$ is the initial velocity at time 0. Then the solution of the differential equation is of type (5.4), the function *distance–time* is $x(t) = x_0 + \dot{x}_0 t + gt^2/2$.

The next theorem is more general than the previous one.

Theorem 5.4. *Equation with separable variables. Let the scalar function f be separable: $f(t, x) = a(t)b(x)$ and have no roots in a certain rectangle. Then the solution of the differential equation (5.1) with the initial condition x_0 (lying in the corresponding interval) satisfies the following implicit equation:*

$$(5.5) \qquad \int_{x(0)}^{x(t)} \frac{1}{b(\xi)} \, d\xi = \int_0^t a(\tau) \, d\tau.$$

Remark. For $b(x) \equiv 1$, (5.5) reduces to (5.4).

Proof. Abusing the congenial system of notations introduced by Leibniz, consider $\dot{x} = dx/dt$ as a normal fraction. After rearrangement, one obtains

$$\frac{dx}{b(x)} = a(t)dt.$$

Integrating both sides between 0 and t (x_0 and $x(t)$, respectively), it yields (5.5). By differentiating the result, one can establish the correctness of the solution if not the method.   ∎

Problem 5.4. Solving differential equations with separable variables. Using Theorem 5.4, 'solve' Examples 5.1, 5.3 and 5.4.

Having seen two approximation methods at work, we will now illustrate a third one. Already Newton knew that it is worthwhile to look for a power series as a solution of a differential equation with analytic R.H.S.

Example 5.8. Solution as power series. Consider the differential equation $\dot{x} = \lambda x$, $x_0 = 1$ (Example 5.1). Let us assume that the solution has the following form: $x(t) = \sum_{k=0}^{\infty} a_k t^k$. Then $\dot{x}(t) = \sum_{k=0}^{\infty} k a_k t^{k-1} = \sum_{k=0}^{\infty} (k+1)a_{k+1} t^k$. Substitute both expressions into the corresponding sides of the equation and identify the coefficients of t^k from the recursive formula: $(k+1)a_{k+1} = \lambda a_k$. With initial condition $x(0) = a_0 = 1$, $a_k = \lambda^k/k!$, $k = 1, 2, \ldots$, yielding the Taylor-series of $e^{\lambda t}$.

Stationary point and stability

For the sake of simplicity, we shall only consider time-invariant systems called *autonomous systems*:

$$(5.6) \qquad\qquad \dot{x} = \frac{dx}{dt} = f(x).$$

In the theory of differential equations the stationary (equilibrium, normal) point plays an outstanding role.

A point $x^\circ \in \mathcal{X}$ is called a *stationary point* of system f if, starting the system from x°, it stays there:

$$(5.7) \qquad \text{if} \quad x_0 = x^\circ, \quad \text{then} \quad x(t) = x^\circ \quad \text{for} \quad t \in \mathbf{T}: \quad f(x^\circ) = 0.$$

We shall repeat the definitions concerning stability of the stationary point discussed in Section 1.1 for difference equations.

1. A stationary point is called *Lyapunov stable* if for any close enough initial state x_0, the resulting trajectory $x(t)$ exists and stays close forever to x°. In formula: for any positive real ε, there exists a positive real δ_ε such that if $||x_0 - x^\circ|| < \delta_\varepsilon$, then $||x(t) - x^\circ|| < \varepsilon$.

2. A Lyapunov stable stationary point x° is called *locally asymptotically stable* if any path $\{x(t)\}$ starting from an initial point x_0 close enough to x°, converges to the stationary point.

3. A Lyapunov stable stationary point is called *globally asymptotically stable* if any path generated by almost any initial point x_0 converges to it. (Of course, if other stationary points exist, they should not be considered for convergence, Example 3.3!)

4. A stationary point is (Lyapunov or asymptotically) *unstable* if and only if it is not (Lyapunov or asymptotically) stable, (respectively).

The next example demonstrates the necessity of requiring Lyapunov stability in the definition of asymptotic stability.

Example 5.9. (See Example 1.4.) Convergence without Lyapunov stability. Consider a planar system given in polar coordinates: $\dot{r} = r(1 - r)$, $\dot{\vartheta} = r\vartheta$ where $r > 0$, $0 \leq \vartheta \leq 2\pi$. It is easy to demonstrate that there exists a unique steady state $(1, 0) = (1, 2\pi)$, attracting all the paths. But regardless of how close the initial state ϑ_0 is to 0, $\dot{\vartheta} > 0$, that is, the system makes a full turn before returning to the steady state.

5.2 LINEAR SYSTEMS

Before continuing the general analysis for nonlinear functions, it is worthwhile considering the simplest case, namely, *linear functions*.

System (5.6) is called *linear* if f is a linear function: there exists an n-dimensional square matrix M and an n-vector w, such that $f(x) = Mx + w$. Then (5.6) simplifies to

$$(5.8) \qquad \dot{x} = Mx + w.$$

We shortly refer to the coordinate form of this system:

$$(5.8') \qquad \dot{x}_i(t) = \sum_{j=1}^{n} m_{ij} x_j(t) + w_i, \qquad i = 1, \ldots, n$$

where $M = (m_{ij})$ is the matrix of transformation M in a fixed coordinate system and $w = (w_1, \ldots, w_n)^{\mathbf{T}}$ is the corresponding ordered n-tuple.

The implicit equation of the stationary point is

$$(5.8°) \qquad\qquad Mx° + w = 0,$$

which, under suitable regularity assumptions, can be made explicit:

$$(5.9) \qquad\qquad x° = -M^{-1}w.$$

The analogy with difference equations would be clearer if (5.8) were replaced by

$$(5.8^*) \qquad\qquad \dot{x} = (M - I)x + w.$$

(see (5.9) and (1.3)). As this procedure would be unusual and only make the formulas cumbersome, we do not use it.

Homogeneous equations

Introduce the *deviation vector*

$$(5.10) \qquad\qquad x^d = x - x°$$

and deduct (5.8°) from (5.8):

$$(5.11^d) \qquad\qquad \dot{x}^d = Mx^d.$$

In words: the deviation vectors satisfy the homogeneous system (5.11^d) obtained from the inhomogeneous system (5.8) by deleting the additive term w.

From now on we shall consider the homogeneous system and for compactness, we shall drop the superscript ^{d}. (We also could say that $w = 0$.) For the sake of reference, we repeat (5.11^d) in this new notation:

$$(5.11) \qquad\qquad \dot{x} = Mx.$$

In the scalar case it is well-known that the solution to (5.11) is $x(t) = e^{Mt}x(0)$ (see Example 5.1). The following question arises: Is it possible to generalize this formula to any matrix M? The answer is yes, but we should first define e^{Mt}. On the basis of Theorem A.6, this could be done, using the appropriate Taylor series:

$$(5.12) \qquad\qquad e^{Mt} = \sum_{k=0}^{\infty} \frac{(Mt)^k}{k!}.$$

Using the theory of matrix-valued power-series, it can be shown that the matrix exponent is a differentiable function of t, and the order of derivation and summation can be interchanged. Hence we obtain the formula (already well-known for scalar M): $d[e^{Mt}]/dt = Me^{Mt}$.

Now we can announce

Theorem 5.5. *The unique solution of the autonomous homogeneous linear system (5.11) with the initial state x_0 can be extended to the whole semi-axis $t \geq 0$ and it is given by*

$$(5.13) \qquad x(t) = e^{Mt} x_0, \quad t \geq 0.$$

We shall calculate now the matrix exponent.

Characteristic roots and characteristic vectors

Similarly to the linear difference equations, it is practically impossible to determine $x(t)$ through the closed-form solution (5.13) for the multivariate linear differential equations. As is known from linear algebra (see Appendix A), with the help of the characteristic vectors and characteristic roots of matrix M, e^{Mt} can be determined relatively easily. Recall (Appendix A) that

$$(5.14) \qquad P(\lambda) = \det(\lambda I - M)$$

is called the *characteristic polynomial* of matrix M. $P(\lambda)$ has n (real or complex) roots with multiplicity, which are the *characteristic roots* of M:

$$(5.15) \qquad P(\lambda_j) = 0, \qquad j = 1, \ldots, n.$$

Let us assume that there exist n linearly independent characteristic vectors, that is, a *characteristic basis*:

$$(5.16) \qquad M s_j = \lambda_j s_j, \qquad j = 1, 2, \ldots, n.$$

With its help, the initial state can be represented as

$$(5.17) \qquad x_0 = \sum_{j=1}^{n} \xi_j s_j.$$

Since $M^t s_j = \lambda_j^t s_j$ and

$$e^z = \sum_{k=0}^{\infty} \frac{z^k}{k!},$$

(5.12), (5.13), (5.16) and (5.17) yield

$$(5.18) \qquad x(t) = \sum_{j=1}^{n} \xi_j e^{\lambda_j t} s_j.$$

We have proved the analogy of Theorem 1.2.

Theorem 5.6. *If M has a characteristic basis, then with the help of the characteristic vectors of M, the initial state can be represented as (5.17), and also using the characteristic roots of M, the t-th state can be represented as (5.18).*

Remarks.   1. Mathematically, the use of characteristic polynomial transforms a first-order n-dimensional system into an n-order scalar system, reversing the transformation described at the beginning of the Section.

2.* For a characteristic vector having characteristic root λ with multiplicity r, the term $e^{\lambda t}$ is accompanied by other terms $t^j e^{\lambda t}$, $j = 1, 2, \ldots, r - 1$.

Complex characteristic roots

Equation (5.18) can be directly used if all the characteristic roots of M are real. What to do, however, if some roots are complex?

Repeating the argumentation of Section 1.2: each characteristic root λ, each characteristic vector s and coefficient ξ also appear with their complex conjugates. Hence

(5.17*) $$x_0 = \xi s + \bar{\xi}\bar{s},$$

(5.18*) $$x(t) = e^{\lambda t}\xi s + e^{\bar{\lambda} t}\bar{\xi}\bar{s}.$$

Taking $\xi = 1$, then by Euler's formula:

(5.19) $$e^{\lambda t} = e^{\Re\lambda t}[\cos\varphi t + i \sin\varphi t], \quad \varphi = \Im\lambda$$

where $\Re z$ and $\Im z$ are real and imaginary parts of a complex number z, respectively.

(5.17*), (5.18*) and (5.19) imply the analogy of Theorem 1.3:

Theorem 5.7. *If matrix M has a simple complex root λ and a complex characteristic vector s, then the corresponding block-solution has the form*

$$x(t) = 2e^{\Re\lambda t}(\Re s \cos\varphi t - \Im s \sin\varphi t).$$

The following problem illustrates Theorem 5.7.

Problem 5.5. (See Example 1.8.) Consider $\dot{x} = Mx$ where

$$M = \begin{pmatrix} 0 & -1 \\ 1 & 0 \end{pmatrix} \quad \text{and} \quad x_0 = \begin{pmatrix} 1 \\ 0 \end{pmatrix}.$$

Prove that $x_1(t) = \cos t$ and $x_2(t) = \sin t$.

The following example shows two methods of solution for higher-order systems.

Example 5.10. Second-order homogeneous linear differential equation. $\ddot{y} + \alpha\dot{y} + \beta y = 0$ can be solved as follows:

a) *Reduction to a first-order planar system.* $x_1 = y$ and $x_2 = \dot{y}$. Then $\dot{x}_1 = \dot{y} = x_2$ and $\dot{x}_2 = \ddot{y} = -\alpha\dot{y} - \beta y = -\alpha x_2 - \beta x_1$. Then $n = 2$,

$$M = \begin{pmatrix} 0 & 1 \\ -\beta & -\alpha \end{pmatrix}.$$

b) *The method of complex amplitudes.* We shall search for a solution in the form $y(t) = y_0 e^{\lambda t}$. Take the derivative of the hypothetical solution, substitute it to the differential equation and divide both sides by $y_0 e^{\lambda t}$: we obtain the quadratic (characteristic) equation $\lambda^2 + \alpha\lambda + \beta = 0$.

In both cases we can continue the solution as described above. For $\alpha = 0$, $\beta > 0$, we obtain a slightly generalized version of Problem 5.5: $y(t) = A\cos(\varphi t + \delta)$ where A, φ and δ are real constants.

Stability and instability

For linear systems, the stability conditions can be given relatively simply.

Theorem 5.8. *Linear stability. The continuous-time linear system (5.11) is stable if and only if $\Re\lambda_j < 0$, $j = 1, \ldots, n$.*

Proof. For simple roots, we can rely on (5.18). For any real characteristic root λ, $\lim_{t \to \infty} e^{t\lambda} = 0$ holds if and only if $\lambda < 0$. For a complex characteristic root, Theorem 5.7 is used. Then $\lim_t x(t) = 0$ if and only if the real part of every characteristic root is negative. For multiple roots, the additional terms $t^j e^{\lambda t}$ converge to zero, too.

Remarks. 1. For continuous-time linear systems, local and global stability are also equivalent.

2. If $\Re\lambda_j \leq 0$ and for example, $\Re\lambda_1 = 0$, then for a simple (pair of conjugate) root(s), Lyapunov stability can be proved. For multiplicity, instability prevails, at least for almost all initial states.

3. Remember Theorem 1.4: for discrete-time systems, a matrix is called stable if all of its characteristic roots lie in the open unit disk.

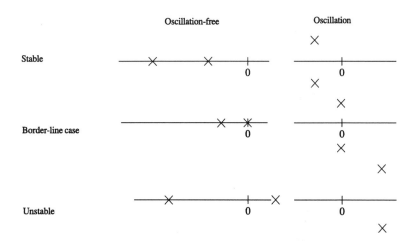

Figure 5.5
Classification of charecteristic roots

In contrast, for continuous-time systems, a matrix is called *stable* if all of its characteristic roots lie in the left half plane (Figure 5.5).

As an analogy of Example 1.12, we have

Problem 5.6. Stability in the plane. The second-order linear differential equation $\ddot{y} + \alpha\dot{y} + \beta y = 0$ is stable if and only if $\alpha,\ \beta > 0$.

Both in the mechanic and economic applications the so-called saddle point plays an important role. We shall only consider the simplest case.

Example 5.11. Saddle point. Let $n = 2$ and one of the characteristic roots of M be negative, the other be positive: $\lambda_1 < 0 < \lambda_2$, with the corresponding characteristic vectors s_1 and s_2. Then paths starting from $x_0 = \xi_1 s_1$ are stable, while those starting from $x_0 = \xi_2 s_2$ are unstable.

5.3 NONLINEAR SYSTEMS

We turn now to the analysis of nonlinear differential equations. First we show an example.

Example 5.12. The nonlinear differential equation $\dot{x} = -x + x^2$ is locally stable but globally unstable. Stationary points: $x^\circ = 0$ and $x^\circ = 1$. With simple calculation it can be established that our nonlinear difference equation converges to 0 for initial states $-\infty < x_0 < 1$ (local stability). In fact, for $x(t) < 0$, $x(t)$ grows, for $0 < x(t) < 1$, $x(t)$ decreases. But for $x(t) > 1$, $x(t)$ also grows, this time to the infinity.

Global stability

We start the discussion with global stability. We shall discuss the famous Lyapunov method (introduced already in Section 3.1). We shall assume the uniqueness of the stationary point x°.

To begin with, we shall define the concept of a *Lyapunov function* with respect to the differential equation (5.6). Assume that the solutions of (5.6) lie in a compact domain. Assume the existence of a function $V(x) : \mathbf{R}^n \to \mathbf{R}$ such that (i) it has a unique global minimum at x° and (ii) $dV[x(t)]/dt < 0$ for every $x(t) \neq x^\circ$.

Remarks. 1. Note that $dV(x)/dt$ is the scalar product of a row and a column vector: $dV(x)/dt = V_x(x)\dot{x}$.

2. It is difficult to find Lyapunov functions in general, but in special cases like in Theorem 5.10 and Section 6.3 there are suitable methods.

Theorem 5.9. (*Lyapunov's theorem on global stability.*) *The nonlinear system (5.6) is globally stable at x° if it has a Lyapunov function.*

Sketch of the proof. We shall assume the contrary, that is, the implication does *not* hold. We shall denote the path starting from x_0 at $t = 0$ by $x(t, x_0)$. Since the solution lies in a compact set, one can select a series of instants $\{t_k\}$, such that the series of states converges to some x^1: $\lim_{k \to \infty} x(t_k, x_0) = x^1 \neq x^\circ$. Because of monotonicity,

$$V[x(t_k, x_0)] > V[x(t_{k+1}, x_0)], \qquad k = 1, 2, \ldots.$$

Because of continuity, $\lim_{k \to \infty} V[x(t_k, x_0)] = V[x^1]$, further,

$$(5.20) \qquad V[x(t_k, x_0)] > V[x^1], \qquad k = 1, 2, \ldots.$$

Let us restart the system from x^1 and select $\{x(u_k, x^1)\}$ which converges to x^2: $\lim_{k \to \infty} x(u_k, x^1) = x^2 \neq x^1$. Similarly to (5.20),

$$(5.21) \qquad V(x^1) > V(x^2).$$

Solutions are characterized by the following property: $x(t+u, x_0) = x[u, x(t, x_0)]$. Consider the series $\{t_k + u_k\}$ assembled from the previously defined two series and its V-image:

$$(5.22) \quad V[x(t_k + u_k, x_0)] = V[x(u_k, x(t_k, x_0))], \qquad k = 1, 2, \ldots.$$

For any k, there exists an integer K such that $t_k + u_k < t_K$, that is,

$$V[x(t_k + u_k, x_0)] > V[x(t_K, x_0)] > V(x^1).$$

On the other hand, there exists another subsequence (not denoted separately) for which $\lim_{k \to \infty} V[x(u_k, x(t_k, x_0))] = V(x^2)$.

As a consequence, for any large enough k, the L.H.S. of (5.22) is larger than $V(x^1)$, the R.H.S. of (5.22) is less than $V(x^1)$, one has a contradiction. ∎

The following example is an excellent illustration for using the Lyapunov function.

Example 5.13. (Hahn, 1967, p. 77.) Consider the planar nonlinear differential equation $\dot{x}_1 = x_2 + \kappa x_1(x_1^2 + x_2^2)$ and $\dot{x}_2 = -x_1 + \kappa x_2(x_1^2 + x_2^2)$. Using the simplest quadratic Lyapunov function $V(x_1, x_2) = (x_1^2 + x_2^2)/2$, we can prove that a) for $\kappa < 0$, it is stable and b) for $\kappa > 0$, it is unstable. Indeed,

$$\frac{dV(x_1, x_2)}{dt} = x_1 \dot{x}_1 + x_2 \dot{x}_2$$
$$= x_1[x_2 + \kappa x_1(x_1^2 + x_2^2)] + x_2[-x_1 + \kappa x_2(x_1^2 + x_2^2)] = 2\kappa V(x_1, x_2)^2.$$

For $\kappa = 0$, we return to the simple cycle of Problem 5.5.

Local stability

We now turn to the issue of local stability. Of course, we assume that there exists at least one stationary point. Let f have a continuous derivative at its stationary point $x^\circ = 0$, hence by (5.7), $0 = f(0)$; and let

$$(5.23) \qquad\qquad M = \mathbf{D}f(0)$$

where $\mathbf{D}$ is the differential operator. Now $\dot{x} = Mx$ [(5.11)] will be called the *linearized part* of (5.6).

Theorem 5.10. *(Perron's theorem on local stability.) The non-linear system (5.6) is locally stable if its linearized part is stable.*

Sketch of the proof. The proof consists of two steps: (i) We show that there is a symmetric positive definite matrix W for which the quadratic form $V(x) = x^{\mathrm{T}}Wx$ is a Lyapunov function for the stable linearized system. (ii) For a sufficiently close neighborhood of 0, $V(x)$ is also a Lyapunov function for the original system.

Ad (i). By a well-known theorem of linear algebra, any matrix M can be transformed into an upper triangular matrix with a unitary transformation S, $SS^{\mathrm{T}} = I$ (Lancaster, 1969, Section 2.10). Then applying transformation $N = S^{-1}MS$, $y = S^{-1}x$, (5.6) is replaced by

$$\dot{y} = Ny + b(y) \quad \text{where} \quad \lim_{y \to 0} \frac{|b(y)|}{|y|} = 0.$$

Moreover, for any $\varepsilon > 0$, there exists a unitary transformation S such that the sum of the moduli of entries above the diagonal of N is less than ε. Suppose that M is already given in such form.

Ad (ii). $V(x) = x^{\mathrm{T}}x$ is a quadratic form and we shall prove that it is a local Lyapunov function. Take the derivative of $V(x)$ with respect to x, plug in (5.6) and arrange the terms:

$$\frac{dV(x)}{dt} = \dot{x}^{\mathrm{T}}x + x^{\mathrm{T}}\dot{x} = [Mx + b(x)]^{\mathrm{T}}x + x^{\mathrm{T}}[Mx + b(x)]$$
$$= x^{\mathrm{T}}M^{\mathrm{T}}x + x^{\mathrm{T}}Mx + b(x)^{\mathrm{T}}x + x^{\mathrm{T}}b(x).$$

Note that $|x^{\mathrm{T}}b(x)| \leq |x|\,|b(x)|$ which is of the order of $o(|x|^2)$. Putting them temporarily aside, taking into account the coefficients of the last equation and the special form of M:

$$x^{\mathbf{T}} M^{\mathbf{T}} x + x^{\mathbf{T}} M x = \sum_i \sum_j (\bar{x}_i m_{ji} x_j + \bar{x}_i m_{ij} x_j)$$

$$< \sum_i \bar{x}_i (\lambda_i + \bar{\lambda}_i) x_i + 2\varepsilon \sum_i \bar{x}_i x_i.$$

By Theorem 5.8, $\lambda_i + \bar{\lambda}_i = 2\Re\lambda_i < 0$ for all i's, hence $dV(x)/dx < 0$ holds, even after taking care of the nonlinear terms. ∎

Remarks. 1.* Note that with this method Theorem 5.8 can be proved without the Jordan normal form.

2. If at least one characteristic root of M has positive real part, then the system is unstable.

3. Example 5.13 illustrates that in the borderline case when the linearized part is Lyapunov stable, the nonlinear system may be asymptotically unstable (for $\kappa > 0$) or asymptotically stable (for $\kappa < 0$).

Cycle and chaos

Until now we have only considered stability of the stationary point. Here we shall only refer to cycles and chaos, thoroughly discussed in the relevant parts of Chapter 3.

Example 5.14. Small vibrations. The equation of the pendulum (Example 5.6) can be linearized for small amplitudes:

$$(5.24) \qquad \ddot{y} = -y, \quad y_0 = A, \quad \dot{y}_0 = 0.$$

For $\beta = 1$ and $\alpha = 0$, Example 5.10 yields solution $y(t) = y_0 \cos t$ and period $P_0 = 2\pi$.

As is known from the history of mechanics, Galilei wrongly supposed that the period of the pendulum is independent of the amplitude. But this claim is a very good approximation. Sophisticated methods show (Arnold, 1973) $P_A = 2\pi[1 + A^2/16 + O(A^4)]$ where $O(A^4)$ is a quantity which remains bounded after dividing by A^4. For example, at an amplitude of 30° the actual period only surpasses the idealized one by 2 per cent!

In the theory of differential equations, the Poincaré–Bendixson theorem on the existence of planar limit cycles is well-known (Coddington and Levinson, 1955, Chapter 16).

The modern theory of chaos has started with a meteorological application of a three-dimensional system of nonlinear differential equations. E. N. Lorenz (1963) noticed that by replacing the original initial state given by many digits by its rounded-off version, the model blew off (for a popular but exciting survey, see E. N. Lorenz, 1993).

5.4 CONTROL IN CONTINUOUS TIME

For the sake of further applications (Chapters 6, 9 and 10), we shall shortly explore continuous-time control systems.

Control systems

While in Section 1.4 we confined our attention to time-invariant control systems with discrete time, now we extend the analysis to time-variant systems with continuous time. The notion of *control system* has a central role in engineering as well as in economic applications. The basic concepts are as follows: *state vector* and *control vector*, denoted by x and u, respectively. The *state space equation* determines the change in the state vector as a function of the time, the state and the control vectors:

$$(5.25) \qquad \dot{x} = g(t, x, u)$$

where function $g : \mathbf{R}^{1+n+m} \to \mathbf{R}^n$.

We speak of *feedback control* if the control only depends on the time as well as on the current state:

$$(5.26) \qquad u = h(t, x)$$

where function $h : \mathbf{R}^{1+n} \to \mathbf{R}^m$.

Substitution of (5.26) into (5.25) yields the so-called
Basic equation:

$$(5.27) \qquad \dot{x} = f(t, x) = g[t, x, h(t, x)].$$

For time-invariant systems, (5.25)–(5.27) reduce respectively to

$$(5.25^*) \qquad \dot{x} = g(x, u),$$
$$(5.26^*), \qquad u = h(x),$$
$$(5.27^*) \qquad \dot{x} = f(x) = g[x, h(x)].$$

In case of time-invariant systems, we speak of *stabilization* if the feedback stabilizes the state space equation, that is, the stationary point of the basic equation is stable. In this case, x° provides a stationary control vector u°:

$$(5.28) \qquad f(x^\circ) = 0 \quad \text{and} \quad u^\circ = h(x^\circ).$$

We copy the definition of a *totally decentralized control* from Section 1.4; the number of control variables is equal to that of state variables: $m = n$; and the feedback is decomposed into n independent scalar feedbacks:

$$(5.29) \qquad u_i = h_i(x_i), \qquad i = 1, \ldots, n.$$

Note, however, that we can also consider as control system the following framework:

$$(5.25') \qquad x = g(u),$$
$$(5.26') \qquad \dot{u} = h(x, u);$$

or in case of decentralized control,

$$(5.29') \qquad \dot{u}_i = h_i(x_i, u_i), \qquad i = 1, \ldots, n.$$

We shall meet such control systems in Section 6.3.

Linear control systems

In the remaining part of this Section, we shall confine our attention to the linear class of control systems (Martos, 1981). First we calculate the solution of

$$(5.30) \qquad \dot{x} = Ax + Bu + p$$

where the initial state x_0 is given.

We shall again use the method of multipliers. Multiplying both sides of (5.30) by e^{-At}:

$$(5.31) \qquad e^{-At}\dot{x}(t) - e^{-At}Ax(t) = e^{-At}[Bu(t) + p].$$

The L.H.S. is equal to the derivative of $e^{-At}x(t)$, that is, by Theorem 5.3, it implies

$$e^{-At}x(t) - x_0 = \int_0^t e^{-A\tau}[Bu(\tau) + p]d\tau.$$

After algebraic manipulation one gets

$$(5.32) \qquad x(t) = e^{At}x_0 + \int_0^t e^{A(t-\tau)}[Bu(\tau) + p]\,d\tau.$$

It is worthwhile mentioning that both terms standing on the R.H.S. of (5.32) have an intuitive meaning: (i) the first term is the homogeneous solution ($u = p = 0$) with the initial state x_0, while (ii) the second term is the inhomogeneous particular solution with the initial state $x_0 = 0$.

We speak of *linear feedback* if the control vector is a linear function of the state vector:

$$(5.33) \qquad\qquad\qquad u = -Kx + q.$$

In Chapter 8 we shall see (although in a discrete-time model) that for quadratic objective function, the optimal control has form (5.33).

6. CONTINUOUS-TIME MODELS

In this Chapter we turn to the discussion of continuous-time economic systems. We shall discuss three fields. Section 6.1 is devoted to the leading models of classical and neoclassical *growth theory*. Section 6.2 surveys the simplest models of *government stabilization*. Finally, Section 6.3 deals with the *walrasian price adjustment process*. Detailed references are given in the Sections but Gandolfo (1971) covers all the three topics.

6.1 GROWTH MODELS

It is well-known that the neoclassical economics hardly paid any attention to the issues of economic growth between 1870 (its date of foundation) and the Second World War. Independently of the discrete-time model of Harrod (1939, see also 1948), Domar (1946, see also 1957) presented a continuous-time model of growth theory. (For a good survey, see Sen, 1970.) These two models, being classical, were only weakly integrated to the mainstream economics. No wonder that the first neoclassical growth model by Solow (1956) attracted a good deal of attention. Since then growth theory has made a great progress (especially now the endogenous growth theory), but we shall only deal with the two pioneering classes of models.

A classical growth model

In Domar's model there are three macro variables: Y = output, I = investment and C = consumption. In continuous-time systems

time dependence is often not denoted. The equations of the model
are as follows.
GDP identity

(6.1) $$Y = C + I.$$

Growth in output

(6.2) $$\dot{Y} = AI, \qquad A > 0.$$

Consumption function

(6.3) $$C = (1 - s)Y, \qquad 0 < s < 1.$$

Both A and s are time-invariant parameters.

We have already discussed a variant of the discrete version of the
present model in Section 4.1.

For better understanding, it is suitable breaking down (6.2) into
two equations:
Constant output-capital ratio

(6.2a) $$Y = AK.$$

Instantaneous capital formation

(6.2b) $$\dot{K} = I.$$

Indeed, taking the derivative of (6.2a) and plugging (6.2b) into it,
(6.2) is obtained.

We have only an equilibrium path now. With a simple substitution, $I = sY$, thus we have the following
Basic differential equation

(6.4) $$\dot{Y} = AsY, \qquad Y(0) = Y_0$$

which is a homogeneous linear equation with a constant coefficient.
On the basis of Example 5.1 we obtain

Theorem 6.1. *(Domar, 1946.) The unique path of the classical growth model (6.4) is exponential*

$$(6.5) \qquad\qquad Y(t) = Y_0 e^{\Gamma t}$$

with a growth rate

$$(6.6) \qquad\qquad \Gamma = As.$$

To give a feeling for real life magnitudes, we mention the following triple of coefficients.

Example 6.1. Numerical illustration. $s = 0.2$, $A = 0.25$, $\Gamma = 0.05$.

To see the other side of the coin, solve

Problem 6.1. Harrod's model. Write down the discrete-time version of the classical growth problem without lags.

Remark. Since its discovery, formula $\Gamma = As$ has provoked a lot of discussions. Let us suppose that the quantity of labor increases by rate ν, its productivity by rate η, then full employment implies $\Gamma = \nu + \eta$. What ensures such an equality? According to the classical theory, such an equality is only an accident and nothing ensures the stability of the equilibrium. Using Harrod's (1939) expression, this equilibrium stands on a knife-edge. This problem is solved by the neoclassical approach.

A neoclassical growth model

The essence of the neoclassical approach lies in the assumption on perfect substitutability between the factors of production. Suppose that output depends on capital as well as labor where $L(t) = L_0 e^{\nu t}$. (Here we neglect the increase in labor productivity: $\eta = 0$.) Using a neoclassical *production function* $Y = F(K, L)$ with constant returns to scale, an arbitrary combination (K, L) yields an output Y. It is worthwhile considering *per capita* output and capital:

$$(6.7) \qquad\qquad y = \frac{Y}{L} \quad \text{and} \quad k = \frac{K}{L}.$$

Then $Y = F(K, L)$ is replaced by

$$(6.8) \qquad\qquad y = f(k) = F(k, 1).$$

Concavity and some regularity conditions are assumed:

(6.9) $f'(\cdot)$ decreases, $f'(0) = \infty$ and $f'(\infty) = 0$.

Retaining (6.1), (6.2b) and (6.3), and the assumption of constant saving ratio, $sf(k)$ is invested at time t. From

$$\left(\frac{\dot{K}}{L}\right) = \frac{\dot{K}L - K\dot{L}}{L^2} = \frac{I}{Y}\frac{Y}{L} - \frac{K}{L}\frac{\dot{L}}{L} = sy - \nu k$$

we can deduce the following nonlinear
Basic differential equation

(6.10) $\dot{k} = sf(k) - \nu k$.

We have now

Theorem 6.2. *(Solow, 1956.) The neoclassical growth model (6.10) has a unique stationary point k° determined by*

(6.11) $sf(k^\circ) = \nu k^\circ; \quad 0 < k^\circ < \infty;$

which is globally stable.

Proof. Substituting $\dot{k} = 0$ into (6.10), yields the implicit equation (6.11). By our concavity assumption, there exists a unique solution k°. Indeed, by (6.9) and the l'Hospital-rule,

$$\lim_{k \to 0}\left[\frac{f(k)}{k}\right] = \lim_{k \to 0} f'(k) = \infty, \quad \lim_{k \to \infty}\left[\frac{f(k)}{k}\right] = \lim_{k \to \infty} f'(k) = 0.$$

By (6.10), $0 < k < k^\circ$ implies $\dot{k} > 0$, $k^\circ < k < \infty$ implies $\dot{k} < 0$, that is, in both cases k converges to k°. The application of the Lyapunov method (Theorem 5.9) makes the proof really precise. ∎

We shall present the simplest family of production functions.

Example 6.2. Cobb–Douglas production functions. For $Y = AK^\alpha L^{1-\alpha}$, $y = Ak^\alpha$ holds, that is, $k^\circ = (sA/\nu)^{1/(1-\alpha)}$.

The basic idea of the proof of Theorem 6.2 is illustrated in Figure 6.1 with $A = 1$; $\alpha = 0.3$; $\nu = 0.05$; $s_1 = 0.2$ and $s_2 = 0.3$. It is to be underlined that higher k° belongs to higher s.

What happens if we give up the equilibrium in Example 6.2?

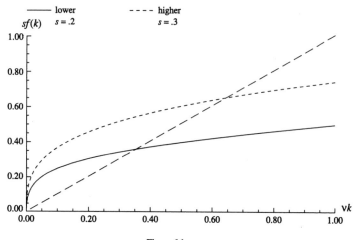

Figure 6.1
Neoclassical growth

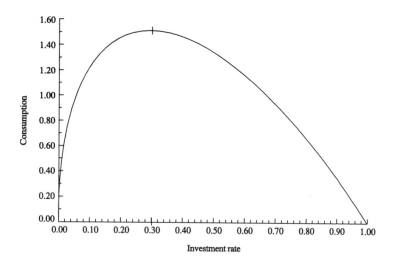

Figure 6.2
Optimal accumulation rate

Problem 6.2. Nonstationary solution (Gandolfo, 1971, 195–6). Determine the nonstationary solution of Example 6.2 by using transformation $z = k^{1-\alpha}$.

In the Introduction, we have quoted Azariadis' (1993) critical opinion just on this model, namely, that Solow set up his model without deriving the behavior equation by optimization. Without accepting this critique, we mention that in Section 10.2 we shall turn to the question of *optimal* accumulation. Here we only mention Phelps (1961) which determined the optimal (golden rule) saving ratio without intertemporal optimization.

Theorem 6.3. *(Phelps, 1961.) The per capita maximal stationary equilibrium consumption is achieved at that capital/labor ratio at which the marginal productivity is equal to the growth rate of labor:*

$$(6.12) \qquad\qquad f'(k^\circ) = \nu.$$

Proof. $C/L = [F(K, L) - \nu K]/L = f(k) - \nu k$ and taking the derivative with respect to k, it yields $f'(k) - \nu$, and so on. ∎

We shall illustrate Theorem 6.3 on Cobb–Douglas production functions.

Example 6.3. Cobb–Douglas optimum. For $y = Ak^\alpha$, (6.12) reduces to $\alpha Ak^{\alpha-1} = \nu$, whence $k^\circ = (\alpha A/\nu)^{1/(1-\alpha)}$. A comparison with Example 6.2 yields $s^\circ = \alpha$.

In Figure 6.2 the optimum is displayed, using the data of Figure 6.1.

Assume that the technological progress is *Solow neutral*, that is, every year the same 'quantity' of capital and labor produces higher output than before: $\mathbf{F}(t, K(t), L(t)) = e^{\eta t} F(K(t), L(t))$. Then using the notation $\Gamma = \nu + \eta$, the so-called *disembodied technological progress* can also be modelled and measured. This was the second great achievement of Solow (Solow, 1957).

6.2* GOVERNMENT STABILIZATION

In the previous Section we neglected a basic idea of Keynes (1936): an active government intervention can stabilize the business fluctuations caused by the private sector of the economy. This idea was very influential in the 1950s and 1960s but it has largely lost its appeal by now.

Following Gandolfo (1971, Section 5.1), we shall present the corresponding stabilization model of Phillips (1954). For simplicity, growth is neglected and the equilibrium values are taken to be zero, thus all variables are *deviations* from the equilibrium.

First we shall describe the spontaneous functioning of the private sphere. We have two equations: one describing the supply, the other the demand.

Aggregate excess demand

$$(6.13) \qquad D = (1-a)Y - u, \quad 0 < a < 1$$

where D is the (aggregate) excess demand of the private sector, a is the marginal rate of spending and u is an exogenous disturbance.

Output response to excess demand

$$(6.14) \qquad \dot{Y} = \alpha(D - Y), \quad \alpha > 0$$

where α is the proportionality factor.

After some calculation with $u = 1$ and $Y(0) = 0$ we have the *Basic equation of the spontaneous behavior*

$$(6.15) \qquad Y(t) = \frac{e^{-\alpha a t} - 1}{a}.$$

By (6.15), $\lim_t Y(t) = -1/a$.

Government stabilization

We turn to the analysis of government stabilization. Following the methods of classical control theory used in technological sciences, Phillips distinguished three stabilization policies.

Proportional stabilization policy where the government spending is proportional to and is of opposite sign of the output tension:

$$(6.16) \qquad G_p^* = -f_p Y, \quad f_p > 0.$$

Derivative stabilization policy where the government spending is proportional to and is of opposite sign of the derivative of output tension:

$$(6.17) \qquad G_d^* = -f_d \dot{Y}, \quad f_d > 0.$$

Integral stabilization policy where the government spending is proportional to and is of opposite sign of the integral of output tension:

$$(6.18) \qquad G_i^* = -f_i \int Y \, dt, \qquad f_i > 0.$$

The resulting government policy is equal to the sum of the three types:

$$(6.19) \qquad G^* = G_p^* + G_d^* + G_i^*.$$

In fact, G^* is merely the *desired* government policy (distinguished by *) from which the actual policy G deviates according to a proportionality factor $\beta > 0$:

$$(6.20) \qquad \dot{G} = \beta(G^* - G), \qquad \beta > 0.$$

Of course, government demand now is added to that of the private sector:

$$(6.13') \qquad D = (1 - a)Y + G - u, \qquad 0 < a < 1.$$

We have completed the stabilization model comprised by (6.13'), (6.14), (6.19), (6.20) and $u = 1$.

Some calculation provides the
Basic differential equation

$$(6.21) \qquad \ddot{Y} + (\alpha a + \beta)\dot{Y} + \alpha\beta a Y - \alpha\beta G^* = -\alpha\beta.$$

Substituting the proper combination of (6.16)–(6.18), it yields the differential equation of the actual control. There exists a conflict between the two goals of stabilization, namely, between the improvement of the stationary value and the dampening of oscillations.

Theorem 6.4. *(Phillips, 1954.) a) The pure proportional policy is not capable to eliminate the drop in incomes and provokes damped oscillations.*

b) The derivative stabilization policy improves on the proportional policy.

c) The integrative stabilization policy is successful if its weight is not too large:

$$(6.22) \qquad f_i < (\alpha a + \beta)a.$$

Sketch of the proof. If $G^* = G_p$, then plugging (6.16) in (6.21) leads to the following differential equation:

$$(6.23) \qquad \ddot{Y} + (\alpha a + \beta)\dot{Y} + \alpha\beta(a + f_p)Y = -\alpha\beta.$$

A stationary state is given by

$$(6.24) \qquad Y^\circ = \frac{-1}{a + f_p}$$

while the characteristic equation of the homogeneous solution is

$$(6.25) \qquad \lambda^2 + (\alpha a + \beta)\lambda + \alpha\beta(a + f_p) = 0.$$

■

6.3 COMPETITIVE PRICE ADJUSTMENT

In the previous two Sections we have analyzed rather simple problems. We now turn to the model of competitive price adjustment where we shall encounter real mathematical difficulties. I shall refer to the corresponding chapters 9, 11 and 12 of the monograph by Arrow and Hahn (1971) but I also acknowledge my debt to Arrow and Hurwicz (1958) and the concise survey of Zalai (1989, Chapter 7).

The equilibrium

Let n be an integer and let $p = (p_1, \ldots, p_n)^{\mathbf{T}}$ be a *price vector* of n dimensions. At this price vector the *market excess demand vector* (for short: *excess demand vector*) is denoted by $z(p) = [z_1(p), \ldots, z_n(p)]^{\mathbf{T}}$. It is assumed that the excess demand function is uniquely determined and, excluding money illusion, it is a zero-degree homogeneous function of prices (and incomes):

$$(6.26) \qquad z(\pi p) \equiv z(p) \quad \text{for every} \quad p \in \mathbf{R}_+^n \quad \text{and} \quad \pi \in \mathbf{R}_+$$

where $\mathbf{R}_+^n$ stands for the nonnegative orthant of $\mathbf{R}^n$. Because of the homogeneity assumption, we may normalize prices, that is, we can take the last product as a *numeraire*: $p_n = 1$, or we can normalize the sum of prices: $\sum_i p_i = 1$, that is, $p^{\mathbf{T}}\mathbf{1} = 1$, that is, p belongs to simplex S_n.

In harmony with the equilibrium approach, it is assumed that the aggregate value of the excess demand is identically zero, that is, *Walras' law* holds:

$$(6.27) \qquad p^{\mathbf{T}}z(p) = 0 \quad \text{for every} \quad p \in \mathbf{R}_+^n.$$

A final technical assumption is that the excess demand function $z(p)$ is *continuous*.

In general equilibrium theory it is important that the market excess demand function be derived from the competition of optimization behavior of consumers and producers (for example, Arrow and Debreu, 1954; Arrow and Hahn, 1971, Chapters 3 and 4). But according to a surprising result by Debreu, Mantel and Sonnenschein (see, for example, Debreu, 1974 and Kirman, 1992), any market excess demand function can be derived from the competition of n optimizing consumers, only assuming that the foregoing function satisfies the conditions above (homogeneity, Walras' law and continuity). In this sense individual optimization does not restrict the class of market excess demand functions.

We speak of *equilibrium price vector* and use the notation p° if the corresponding excess demand vector is nonpositive:

$$(6.28a) \qquad z(p^{\circ}) \le 0.$$

Relying on Walras's law, it is easy to show that any free good has zero price:

$$(6.28b) \qquad \text{if} \quad z_i(p^{\circ}) < 0, \quad \text{then} \quad p_i^{\circ} = 0.$$

First we state the existence of an equilibrium price vector.

Theorem 6.5. *(Arrow and Debreu, 1954, Arrow and Hahn, 1971, Theorem 2.2.) Under standard assumptions, there exists at least one equilibrium price vector.*

Sketch of the proof. Because of the forthcoming argumentation, it is worthwhile presenting the basic idea of the proof. We are looking for a mapping T which improves the errors of a nonequilibrium price vector. Let d_i be a positive number and

$$(6.29) \qquad M_i(p) = d_i z_i(p)_+, \qquad i = 1, 2, \ldots, n$$

where x_+ is the positive part of x. Let p^* be the new price vector:

(6.30)
$$p^* = T(p) = \frac{p + M(p)}{[p + M(p)]^{\mathbf{T}} \mathbf{1}}.$$

It is easy to see that T is a mapping of S_n into S_n. By Brouwer fixed-point theorem (Theorem 3.1), T has a fixed point. A simple calculation establishes that any fixed point of T is an equilibrium. ∎

We turn now to the discussion of the uniqueness of equilibrium. We shall only mention one uniqueness theorem which is also used in the proof of stability. We have to work with smooth excess demand functions, otherwise even local uniqueness breaks down.

For differentiable functions, we call two goods ($i \neq j$) *gross substitutes* at price vector p if an increase in the price p_j implies an increase in the excess demand z_i. The adjective *gross* refers to the fact that we have in mind uncompensated (Marshallian) rather than compensated (Hicksian) demand.

It is said that the demand function $z(p)$ satisfies the *weak axiom of revealed preference* if for any $p \neq \pi p^o$ ($\pi > 0$), $p^{o\mathbf{T}} z(p) \geq 0$. It is easy to see that (i) for a single consumer, the excess demand function satisfies the foregoing axiom and (ii) then the proportions of the equilibrium price vector are unique.

Problem 6.3. Prove assertions (i) and (ii).

At this point we jump to the realm of Part II and derive the demand functions from optimization.

Example 6.4. Gross substitutes. (Arrow and Hahn, 1971, p. 225.) For consumers with Cobb–Douglas utility functions, any two goods are gross substitutes at any price vector. To see this, let $u_h(x_1, \ldots, x_n) = \prod_{j=1}^n x_j^{\alpha_{hj}}$ where $\alpha_{hj} > 0$, $\sum_{j=1}^n \alpha_{hj} = 1$. Let $a^h > 0$ be the endowment vector of consumer h. A simple calculation establishes $z_j^h(p) = \alpha_{hj} p^{\mathbf{T}} a^h / p_j$, that is, using notation $z_{jk}^o = \partial z_j(p^o)/\partial p_k$, $z_{jk}^o = \sum_h \alpha_{hj} a_j^h / p_k > 0$ holds if $j \neq k$.

We can now present the uniqueness theorem.

Theorem 6.6. (*Arrow and Hahn, 1971, Corollary of Theorem 9.7.*) *If all goods are gross substitutes at all normalized price vectors, then the normalized equilibrium price vector is unique.*

Proof. Let $p^o = (p_1^o, \ldots, p_n^o)^{\mathbf{T}}$ be an equilibrium price vector. Take another price vector $p = (p_1, \ldots, p_n)^{\mathbf{T}} > 0$. We shall prove that the latter is not an equilibrium. Let $v_i = p_i/p_i^o$ and k be index of the maximum and λ be the value of maximum: $v_k = \max_i(v_i) = \lambda$: $p \leq \lambda p^o$ and at coordinate k equality holds. Due to the homogeneity property and the equilibrium, $0 = z_k(p^o) = z_k(\lambda p^o)$. Consider that decreasing sequence of price vectors with first member λp^o, with last member p, and only the kth component of the kth vector is changing. Because of gross substitutability, then z_k increases at every time, hence $z_k(p) > 0$, that is, p is not an equilibrium. ∎

The theorem have an interesting implication:

Theorem 6.7. *(Arrow and Hahn, 1971, Theorem 9.8.) Let $p > 0$ and p^* be two different price vectors, $v_i = p_i^*/p_i$ and let k be the index of the maximum. Assume that all goods are gross substitutes at all normalized price vectors. Then jumping from p to p^*, the excess demand for good k increases.*

Price adjustment process

We have seen that in the static model there exists an equilibrium and it is often unique. At the end of 19th century, Walras (1874, 1877) and his early followers did not possess the tools of modern general equilibrium theory. Therefore, in addition to the primitive method of comparing the number of equations and variables, they tried to find a heuristic proof which can be considered as a model of market adjustment.

Following Walras, Samuelson (1941, 1947) modelled the following price adjustment process. At instant t, the *auctioneer* announces the actual price vector $p(t)$. At the market, the excess demand vector $z[p(t)]$ is immediately determined. If there is an excess demand on market k at instant t, then the auctioneer increases the price of the good in proportion of the excess demand. In case of excess supply, the opposite rule applies. With the help of the new price, the 'step' is continuously repeated. If the adjustment process is stable, then the price vector obviously converges to the equilibrium. Formally:

$$(6.31) \qquad \dot{p}_i = d_i z_i(p), \qquad i = 1, \ldots, n$$

where $d_i > 0$ is the reaction coefficient of good i, given exogenously. In vector form:

$$(6.32) \qquad \dot{p} = \langle d \rangle z(p)$$

where $\langle d \rangle$ is a diagonal matrix formed from vector d.

Note the similarity between relation (6.29) used in proving Theorem 6.5 and the present (6.31). It is an essential restriction that trade at false prices (before reaching the equilibrium) is forbidden. That is the reason that we speak of *groping*, or in French: *tatonnement*. In the opposite case the endowment vectors as well as the demand functions would change in time. (An alternative assumption states that the goods are perishable, thus there is no accumulation.)

The price adjustment process may lead to prices with negative components if the price of good i drops to zero, but the excess demand for it is still negative. Therefore we have to impose the following modification:

$$(6.33) \quad \dot{p}_i = \begin{cases} d_i z_i(p) & \text{if } p_i > 0 \text{ or } z_i(p) \geq 0; \\ 0 & \text{if } p_i = 0 \text{ and } z_i(p) < 0; \end{cases} \qquad i = 1, \ldots, n.$$

Note that the R.H.S. of the differential equation (6.33) is not continuous, thus the usual theorems of existence and uniqueness (Theorems 5.1 and 5.2) do not apply. We simply assume that the system is well behaved.

Local stability

In contrast to the economic tradition, we shall start the analysis with local stability. Then it is sufficient to consider the version (6.31) or (6.32), but we assume that good n is the numeraire:

$$(6.31') \qquad \dot{p}_i = d_i z_i(p), \qquad i = 1, \ldots, n-1, \quad \text{and} \quad p_n = 1.$$

On the basis of Theorems 5.8 and 5.10, we have

Theorem 6.8. *(Samuelson, 1947, Chapter IX. and Arrow and Hahn, 1971.) The price adjustment process (6.31') is locally stable around p° if the $(n-1) \times (n-1)$ truncated matrix $\langle d \rangle z'(p^\circ)$ is stable.*

We shall now specify Theorem 6.8 into

Theorem 6.9. *(Metzler, 1945.) If all goods are gross substitutes at the equilibrium point p°, then the price adjustment process is locally stable around p° for any reaction coefficient vector d (total stability).*

Sketch of the proof. Let $z_{ij}^o = \partial z_i(p^o)/\partial p_j$, again. Applying Euler's theorem to the zero-degree homogeneous function $z_i(p)$ and rearranging: $\sum_{j=1}^{n-1} p_j^o z_{ij}^o = -z_{in}^o$, $i = 1, \ldots, n-1$.

On the basis of our assumption, $z_{ij}^o > 0$, $i, j = 1, \ldots, n-1$, that is, $\sum_{j=1}^{n-1} p_j^o z_{ij}^o < 0$, $i = 1, \ldots, n-1$. This is equivalent to the stability of the truncated matrix (Theorems 5.8 and A.10). ∎

Global stability

We shall start the analysis of global stability with boundedness.

Theorem 6.10. *(Arrow and Hahn, 1971, Theorem 12.2.) The path of the price adjustment process (6.33) is bounded.*

Proof. We shall prove that the price vector moves on the surface of the following ellipsoid:

$$(6.34) \qquad \sum_{i=1}^{n} \frac{p_i^2}{d_i} = \sum_{i=1}^{n} \frac{p_i^2(0)}{d_i}.$$

Indeed, take the total derivative of the expression on the L.H.S. of (6.34) with respect to time, substitute (6.33) and Walras' law:

$$(6.35) \qquad \sum_{i=1}^{n} \frac{2p_i \dot{p}_i}{d_i} = 2 \sum_{i=1}^{n} p_i z_i = 0.$$

Note that apparently we have only considered the upper branch of (6.33), but considering the lower branch, the same is true, since $p_i = 0$. ∎

We turn now to the study of global stability. We shall start with the simplest case as an exercise.

Problem 6.4. Global stability in the plane (Arrow and Hahn, 1971, Theorem 12.1.) Prove that in a two-commodity economy if equilibrium exists, it is globally stable.

The following theorem is perhaps the most popular in this field.

Theorem 6.11. *(Arrow et al., 1959.) If the excess demand function satisfies the weak revealed preference axiom at every price vector p, then the price adjustment process (6.33) is globally stable around p^o for any reaction coefficient d.*

Problem 6.5. Prove Theorem 6.11 with the Lyapunov method.

We shall present a generalization of Theorem 6.11.

Theorem 6.12. *(Arrow et al., 1959; Arrow and Hahn, 1971, Theorem 12.4.) If all goods are gross substitutes at every point p, then the price adjustment process (6.33) is globally stable around p° for any reaction coefficient $d > 0$.*

Proof. $V(p) = \max_i(p_i/p_i^\circ)$. Using Theorem 6.7, if the excess demand for a good with maximum price increases, its price decreases, V decreases and so on. ∎

Instability

Around 1958 the above mentioned fact, that every conceivable market excess demand function can be derived from individual optimization, was unknown. It was conjectured that every market excess demand function derived from individual optimization is well-behaved and the corresponding price adjustment process is stable. It was really surprising when Scarf (1960) and Gale (1963) came up with counterexamples, showing instability.

(Counter)example 6.5. Instability (Scarf, 1960.) Let the number of goods (indexed by i) and of households (indexed by h) be three: $n = 3$ and $m = 3$. Let the endowment of household h be $w^h = e_h$ (where e_h is the hth unit vector) and the corresponding utility function be $U^h(x) = \min[x_h, x_{h+1}]$, $h = 1, 2, 3$ where $x_4 = x_1$. For a given price vector p, the optimal demand of household h for good i is equal to $x_i^h = p_h/(p_h + p_{h+1})$, $i = h, h+1$ and $x_{h-1}^h = 0$ where $x_0 = x_3$. The resulting market excess demand function is

$$z_i = \frac{p_i(p_{i-1} - p_{i+1})}{(p_i + p_{i-1})(p_i + p_{i+1})}.$$

The equilibrium vector is $p^\circ = 1$. Let the adjustment vector be $d = 3 \cdot 1$. By (6.34), the price vector moves on a ball of radius $\sqrt{3}$. Assume that the initial vector satisfies $p_1(0)^2 + p_2(0)^2 + p_3(0)^2 = 3$, but $p_1(0)p_2(0)p_3(0) \neq 1$. With simple calculation, $d(p_1 p_2 p_3)/dt = 0$, hence there is no convergence. ∎

Adjustment in discrete time*

At the end of the Section and the Chapter, we consider price adjustment in discrete time, a topic belonging to Chapter 4. The situation is much more complex now. Take the following variant of (6.33):

$$(6.36) \qquad p_{i,t+1} = [p_{i,t} + d_i z_i(p_t)]_+, \qquad i = 1, \ldots, n.$$

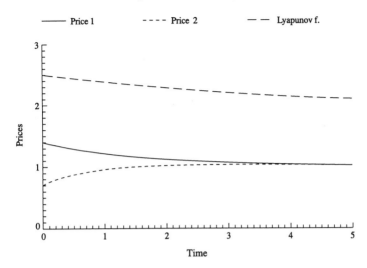

Figure 6.3a
Convergence in the time domain

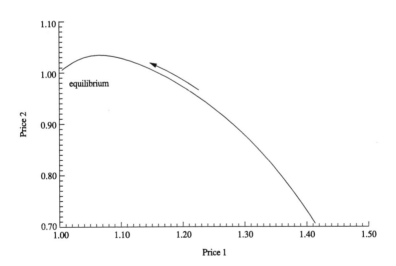

Figure 6.3b
Convergence in the phase plane

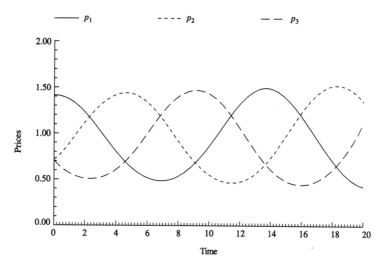

Figure 6.4
A cyclic price adjustment

Using the approximation $dt = 0.1$, Figure 6.3a and b illustrate Example 6.4, $p_1(0) = p_2(0) = \sqrt{2}$ and $p_3(t) \equiv 1$. Note that the third curve in Figure 6.3a representing the Euclidean distance $|p_t - p^o|$, decreases as a Lyapunov function. Figure 6.4 illustrates Example 6.5.

While in continuous-time models only the proportion of the re-action coefficients matters, now their moduli are also important. In the former case, if years were replaced by months, then the intensity of the excess demand would remain the same, but the reaction coefficients should/may be magnified by 12. In the latter case, the excess demand drops to its $1/12$, thus the reaction coefficients should be magnified by 12. Applying the method of Theorem 4.5, we can prove the discrete-time versions of Theorems 6.4 and 6.9.

Theorem 6.13. *If gross substitution holds at the equilibrium point p^o, then the discrete-time price adjustment process (6.36) is locally stable around p^o for any damped reaction coefficient vector d: $0 < d \leq 1$.*

For modern results, see Saari (1985).

Complements

This book contains materials concerning neither *limit cycles* nor *chaos in continuous-time models*. Outstanding contributions to these topics are Goodwin (1951), (1967) and Medio (1991), respectively, see also H.-W. Lorenz (1989).

PART II

DYNAMICS WITH OPTIMIZATION

As was argued in the Introduction, according to the dominant economic theory, the behavior rules of the actors should be derived from optimization. In this Part we shall study the dynamic optimization.

Chapters 7 and 8 return to the discrete-time systems of Chapters 1–4.

Chapter 7 outlines the theory of *dynamic programming* where a time-sum of the reward functions of the state and control vectors of a control system should be maximized with an indirect method.

Chapter 8 *applies* dynamic programming to concrete economic problems.

Chapters 9 and 10 switch to the continuous-time systems of Chapters 5 and 6.

Chapter 9 outlines the theory of *dynamic control* where a time-integral of the reward function of the state and control paths of a control system should be maximized with a direct method.

Chapter 10 applies the foregoing theory to characterize the *optimal consumption paths* with exogenous and endogenous wage and interest dynamics, respectively.

7. DYNAMIC PROGRAMMING

We return to the *discrete-time* systems and models. This Chapter deals with a special version of dynamic optimization, namely, *Dynamic Programming*. The structure of the Chapter is as follows. Sections 7.1 and 7.2 outline the deterministic and stochastic *principle of optimality*, respectively. The remaining Sections consider the so-called *Linear-Quadratic-Gaussian, LQG control problems*. Section 7.3 is devoted to the case of *full information*: LQ control. Section 7.4* refers verbally to the optimal state-estimation and control of the stochastically disturbed systems, the so-called *Kalman-filter*. Dynamic Programming is discussed in much detail by Sargent (1987) and in particular, Stokey and Lucas (1989). A good monograph on optimal LQG control is Bryson and Ho (1969). Our treatment is based on the last two sources.

7.1 DETERMINISTIC PRINCIPLE OF OPTIMALITY

Basic problem

The basic problem of dynamic programming is to maximize a time-additive objective function of a controlled dynamical system.

We shall need the following notations. Time is denoted by $t = 0, 1, 2, \ldots$, the state and control vectors are x_t (n-vector) and u_t (m-vector), respectively. The dynamics of the system is given by the difference equation $x_{t+1} = g_t(x_t, u_t)$ rather than $x_t = g_t(x_{t-1}, u_t)$ of Chapter 1. To evaluate various paths, we introduce the time-additive

157

objective function. Let $f_t : \mathbf{R}^{n+m} \to \mathbf{R}$ be the *return function* of the
state x_t and control u_t, $f_t(x_t, u_t)$, $t = 0, 1, 2, \ldots$.

The basic problem of dynamic programming can be formalized
as follows:

$$(7.1) \qquad f_{T+1}(x_{T+1}) + \sum_{t=0}^{T} f_t(x_t, u_t) \to \max$$

subject to

$$(7.2) \quad x_{t+1} = g_t(x_t, u_t), \qquad t = 0, 1, \ldots, T \quad \text{and} \quad x_0 \quad \text{given.}$$

Remarks. 1. We shall frequently work with the negative of
the return function, called *loss function* and then we shall minimize
rather than maximize the objective function.

2. Term $f_{T+1}(x_{T+1})$, appearing in (7.1), makes it possible that
x_{T+1} be free. If we want, however, x_{T+1} to be fixed, we can modify
f_{T+1} appropriately (see Problem 7.2).

3. We mostly assume that the return functions are increasing,
thus the difference between equality versus inequality constrainsts will
be neglected.

We shall call a dynamic optimization problem *reducible* if (i) the
dimensions of the state and control vectors are the same: $n = m$, and
(ii) the control vector is determined by the neighboring state vectors
via the state space equation: $u_t = \varphi_t(x_t, x_{t+1})$. Substituting into the
utility function and constraint $u_t \in \mathcal{U} \subseteq \mathbf{R}^n$, yield $\psi_t(x_t, x_{t+1}) =$
$f_t[x_t, \varphi(x_t, x_{t+1})]$ and $\varphi_t(x_t, x_{t+1}) \in \mathcal{U}$, respectively. Then the basic
problem of dynamic programming can be described without referring
to the control vectors (Stokey and Lucas confine their attention to
this case, adding time-invarince): Let $\psi_t : \mathbf{R}^{2n} \to \mathbf{R}$ be the *derived
return function*, let $\mathcal{X}$ be a subset of $\mathbf{R}^n$ and let Γ_t be a mapping of
$\mathcal{X}$ into itself. Then

$$(7.1') \qquad \sum_{t=0}^{T} \psi_t(x_t, x_{t+1}) \to \max$$

subject to

$$(7.2') \quad x_{t+1} \in \Gamma_t(x_t), \qquad t = 0, 1, \ldots, T \quad \text{and} \quad x_0 \quad \text{given.}$$

At this point we touch on the following statement.

Lemma 7.1. *Envelope-theorem. Let $F(x,a)$ be a smooth function $\mathbf{R}^{n+1} \to \mathbf{R}$ defined on an appropriate rectangle $I \times J$, where x is a vector variable and a is a parameter. Assume that for every a, function $F(x,a)$ has a unique local maximum, denoted by $x(a)$. Then the maximum value $M(a) = F[x(a),a]$ is also a smooth function and its derivative is $M'(a) = F_a[x(a),a]$.*

Remark. The proposition states that the change in the maximum function with respect to the parameter only depends on the change in the maximand with respect to the parameter.

Proof. Take the total derivative of $M(a)$ with respect to a: $M'(a) = F_x[x(a),a]x'(a) + F_a[x(a),a]$. Since $x(a)$ is a local extremum, $F_x[x(a),a] = 0$, thus the proposition follows. ∎

Direct optimization

Before discussing the standard procedure of dynamic programming, we present the result of a direct optimization, still making use of the special structure of the objective function and the state space equation. We shall need the following notations. $\{p_t\}_0^{T+1}$ is an n-dimensional vector sequence with $p_{T+1} = 0$ and

$$H_t(x_t, u_t, p_t) = f_t(x_t, u_t) + p_t^{\mathbf{T}} g_t(x_t, u_t)$$

is the series of *Hamiltonians* of time-period t, $f_{T+1}(x_{T+1}, u_{T+1}) \equiv f_{T+1}(x_{T+1})$ and $\dfrac{\partial H}{\partial z}$ denotes the row q-vector with entry i being $\dfrac{\partial H}{\partial z_i}$, $i = 1, \ldots, q$.

Theorem 7.1. *Under appropriate smoothness conditions, the optimal solution to (7.1)–(7.2) (if it exists) satisfies the state space equation (7.2),*
the multiplier difference equation (reversed in time)

$$p_{t-1} = \frac{\partial H_t}{\partial x_t}(x_t, u_t, p_t), \qquad t = T+1, \ldots, 1, \qquad p_{T+1} = 0;$$

and the optimality condition

$$\frac{\partial H_t}{\partial u_t}(x_t, u_t, p_t) = 0, \qquad t = T, \ldots, 0.$$

Proof. We set out from the Lagrange function of the conditional maximum problem:

$$L(x, u, p) = f_{T+1}(x_{T+1}) + \sum_{t=0}^{T} \left\{ f_t(x_t, u_t) - p_t^{\mathrm{T}}[x_{t+1} - g_t(x_t, u_t)] \right\}.$$

Calculating the partial derivatives with respect to x_t and u_t, and introducing the Hamiltonian, the two systems of equations are obtained. ∎

Although the system of equations is rather simple, its solution may be simplified by a roundabout method.

The principle of optimality

We turn to the standard problem of dynamic programming. Instead of a single problem we shall consider a sequence of nested problems running from t to $T + 1$, $(t = 0, \ldots, T)$:

$$f_{T+1}(x_{T+1}) + \sum_{\tau=t}^{T} f_\tau(x_\tau, u_\tau) \to \max$$

subject to (7.2), $\tau = t, \ldots, T$ and x_t 'given'.

Assume that each problem has an optimal solution. Let the maximum of problem t be $v_t(x_t)$ as a function of the "initial state" x_t, $t = 0, 1, \ldots, T$. These function are called *value functions*.

It is evident that the following recursive relation holds between subsequent problems. If the optimal solution of problem $t + 1$ with initial condition x_{t+1}° is $u_{t+1}^\circ, x_{t+2}^\circ, \ldots, u_T^\circ, x_{T+1}^\circ$ and the corresponding value function is $v_{t+1}(x_{t+1}^\circ)$, then the optimal solution of problem t with initial condition x_t° is $u_t^\circ, x_{t+1}^\circ, u_{t+1}^\circ, x_{t+2}^\circ, \ldots, u_T^\circ, x_{T+1}^\circ$; and the value function is $v_t(x_t^\circ)$ where u_t° and $v_t(x_t^\circ)$ satisfy the following relation:

$$(7.4) \quad v_t(x_t^\circ) = \max\{f_t(x_t^\circ, u_t) + v_{t+1}(x_{t+1}^\circ) \mid u_t, \quad x_{t+1}^\circ = g_t(x_t^\circ, u_t)\}.$$

Indeed, if u_t° takes the system from x_t° to x_{t+1}°, then we can proceed along the optimum of problem $t + 1$.

It will be useful to illustrate the principle of optimality on an everyday example. Suppose that we want to travel from city A to city B and the optimal path goes through city C. Then the optimum path from city C to city B coincides with the corresponding part of the previous optimum.

At the solution, it is worthwhile proceeding in reversed time. First solve problem (7.4) for $t = T$:

$$f_{T+1}(x_{T+1}) + f_T(x_T, u_T) \to \max \text{ s.t. } x_{T+1} = g_T(x_T, u_T).$$

There exists an optimal control $u_T^\circ = h_T(x_T)$ as a function of state x_T (called *feedback*), which yields *optimal transition* $x_{T+1}^\circ = \phi_T(x_T) = g_T[x_T, h_T(x_T)]$; further, $v_T(x_T) = f_{T+1}[\phi_T(x_T)] + f_T[x_T, h_T(x_T)]$.

By mathematical induction, before turning to problem t, we have already solved problems $T, \ldots, t+1$, knowing $u_\tau = h_\tau(x_\tau)$, $v_\tau(x_\tau)$, $\tau = T, \ldots, t+1$.

Using the recursion, the optimal control $u_t^\circ = h_t(x_t^\circ)$ satisfies the following equation:

$$(7.5) \qquad v_t(x_t^\circ) = f_t[x_t^\circ, h_t(x_t^\circ)] + v_{t+1}\{g_t[x_t^\circ, h_t(x_t^\circ)]\}$$

and maximizes $f_t(x_t^\circ, u_t) + v_{t+1}[g_t(x_t^\circ, u_t)]$ with respect to u_t.

At the end, we solve problem 0 with the *true* initial condition x_0. This procedure gives the principle of optimality of Bellman (1957), which is summarized as

Theorem 7.2. *(Bellman, 1957.) The conditional optimum of problem (7.1) subject to (7.2) can be solved with the following reversed recursion. We solve the simple optimum problem (7.4) for $t = T, T-1, \ldots, 0$. The optimal (feedback) control is $u_t^\circ = h_t(x_t^\circ)$ and the optimal transition is $x_{t+1}^\circ = g_t[x_t^\circ, h_t(x_t^\circ)] = \phi_t(x_t^\circ)$.*

The description above is too abstract and we need more help in applications. Assume that both the state space equations and the return functions are differentiable. Then using the ordinary first-order conditions, we arrive at

Theorem 7.3. *Under suitable smoothness conditions, the necessary conditions of optimum are as follows:*

$$(7.6) \qquad \frac{\partial f_t}{\partial u_t}[x_t, h_t(x_t)] + \frac{\partial v_{t+1}}{\partial x_{t+1}}\{g_t[x_t, h_t(x_t)]\}\frac{\partial g_t}{\partial u_t}[x_t, h_t(x_t)] = 0,$$
$$t = T, T-1, \ldots, 0.$$

Proof. One can prove that the optimal feedback functions h_t are differentiable (Benveniste and Scheinkman: Stokey and Lucas (1989, p. 84)). Then taking the derivative of the R.H.S. of (7.5) yields (7.6). ∎

Remark. In special cases, equation (7.6) can be solved explicitly, but in general, numerical methods are needed.

First we consider the simplest cases.

Example 7.1. Elementary deterministic example. $T = 1$, $n = 1 = m$, dynamics: $x_1 = x_0 + u_0$ and $x_2 = x_1 - u_1$; loss function: $x_1^2 + x_2^2$. $v_1(x_1) = \min\{x_2^2|\ u_1\} = \min\{(x_1 - u_1)^2|\ u_1\} = 0$ if $u_1 = x_1$. $v_0(x_0) = \min\{x_1^2|\ u_0\} = \min\{(x_0 + u_0)^2|\ u_0\} = 0$ if $u_0 = -x_0$.

Problem 7.1. Elementary deterministic problem. $T = 1, n = 1$, state space equation: $x_1 = x_0 + u_0$ and $x_2 = x_1 + u_1$; loss function: $x_0^2 + u_0^2 + x_1^2 + u_1^2 + x_2^2$. Calculate the optimum solution.

Remark. Obviously one can solve Example 7.1 and Problem 7.1 without dynamic programming. For $T \gg 1$, however, dynamic programming is helpful, as the following example shows.

Example 7.2. Special case of Theorem 7.8 below. Let A_t and B_t be two scalar sequences, x_t and u_t scalar state and control variables, respectively and let the state space equation be $x_{t+1} = A_t x_t + B_t u_t$. Finally we present the per-period loss function: $f_t(x_t, u_t) = F_t x_t^2 + G_t u_t^2$ where F_t and G_t are positive reals. It can be proved that the optimal control is $u_t^o = -K_t x_t^o$ where K_t is the time-variant feedback coefficient:

$$K_t = \frac{B_t S_{t+1} A_t}{G_t + B_t^2 S_{t+1}}, \qquad S_t = \frac{G_t A_t^2 S_{t+1}}{G_t + B_t^2 S_{t+1}} + F_t, \qquad S_{T+1} = F_{T+1}.$$

Note that recursion on S_t runs in reversed direction. The solution is illustrated in Figure 7.1 for constant parameters ($A = 2$, $B = 1$, $F = 1$, $G = 5$) and $T = 5$. Note that the short term solution is 'unstable', not approaching zero.

Problem 7.2. Using Example 7.2, prove that if $u_0, \ldots, u_{T-1}$ are positive reals and their sum is unity: $\sum_t u_t = 1$, then the sum of their squares $\sum_t u_t^2$ is minimal if and only if $u_0 = \cdots = u_{T-1} = 1/T$. (If $x_T = 0$, then $F_T = 0$, $A_{T-1} - B_{T-1} K_{T-1} = 0$.) This problem can be solved much more simply without dynamic programming, but it still illustrates the use of dynamic programming.

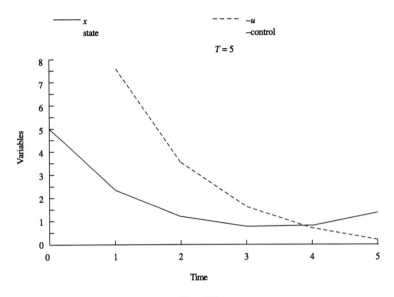

Figure 7.1
Short term optimal LQ control

Time-invariant parameters, discounted objective functions

The basic problem of dynamic programming can be significantly simplified if, apart from discounting, the return functions are time-invariant, and the state space equations are also time-invariant.

$$f_t(x_t, u_t) = \beta^t f(x_t, u_t), \qquad t = 0, 1, \dots, T;$$
$$x_{t+1} = g(x_t, u_t), \qquad t = 0, 1, \dots, T \quad \text{and} \quad x_0 \quad \text{given};$$
$$\beta^{T+1} f_o(x_{T+1}) + \sum_{t=0}^{T} \beta^t f(x_t, u_t) \to \max.$$

Remark. In continuous-time models (Chapter 9) the expres-
sion $e^{-\beta t}$, $\beta \geq 0$ will immediately show that the multiplier is less than
or equal to 1. In discrete-time models this distinction is not usual,

leading to the strange practice that stronger discounting is represented by lower (rather than higher) discount factor.

To simplify the formulas, we introduce the *present value function*:

$$V_t(x_t) = \beta^{-t} v_t(x_t).$$

Dropping the mark of optimum ($^\circ$), then (7.4) simplifies into

(7.7) $V_t(x_t) = \max\{f(x_t, u_t) + \beta V_{t+1}(x_{t+1})| \ u_t, \quad x_{t+1} = g(x_t, u_t)\}.$

Infinite time horizon

Our dynamic programming problem is further simplified if the time horizon is infinite ($T = \infty$):

(7.8) $$\sum_{t=0}^{\infty} \beta^t f(x_t, u_t) \to \max, \qquad 0 < \beta < 1.$$

A further simplification is possible if the time index is dropped and the new state is denoted by x^* in (7.7).

Under suitable assumptions, the infinite series (7.8) is convergent and the optimization problem is meaningful.

Theorem 7.4. *(See Stokey and Lucas, 1989.) Assume that functions f and g satisfy concavity and compactness conditions. Then the optimal feedback function is time-invariant, u = h(x) and, together with the value function V, satisfies the following functional equation:*

(7.9) $V(x) = \max\{f(x, u) + \beta V(x^*)| \ u, \quad x^* = g(x, u)\}.$

Remark. For the infinite horizon problem the end point condition is replaced by the *transversality condition*: $\lim_{t \to \infty} \beta^t V(x_t) = 0$. Its role can be understood at the following rearrangement:

$$V(x_0) = f(x_0, u_0) + \beta V(x_1) = \cdots = \sum_{t=0}^{T} \beta^t f(x_t, u_t) + \beta^{T+1} V(x_{T+1}).$$

The tranversality condition guarantees the asymptotic elimination of the remainder term.

Detour on functional equations

In equation (7.9) a whole function V is the unknown. In such a case we speak of *functional equation*. (As a matter of fact, differential equations discussed in Chapter 5 were already functional equations.)

The next example and problem deal with the simplest functional equation.

Example 7.3. Cauchy-equation. There is a single differentiable (or continuous) solution to the functional equation $v(x + y) = v(x) + v(y)$ for any real pair x, y, and $v(1) = 1$: $v(x) = x$.

As a start, $v(0) = 2v(0) = 0$. a) In case of differentiability the difference quotient $[v(x + y) - v(x)]/y = v(y)/y = [v(y) - v(0)]/y$ converges to $v'(x) = v'(0)$ as $y \to 0$: $v(x) = x$. b) In case of continuity, let n be a positive integer. Then by mathematical induction, $v(nx) = nv(x)$. Let m be also a positive integer; then using the previous relation: $1 = v(1) = mv(1/m)$, hence $v(n/m) = nv(1/m) = n/m$. By continuity, for any irrational x, $v(x) = x$.

Problem 7.3. Characterization of logarithm. There is a single differentiable (or continuous) solution to the functional equation $V(\xi\eta) = V(\xi) + V(\eta)$ for any positive pair ξ, η and $V(e) = 1$ where e is the base of the natural logarithm: $V(\xi) = \log \xi$.

Successive approximation*

If the value function is known, we can determine the optimal feedback via (7.9). In general, however, the value function is not known, either, thus the following procedure is interesting.

The method of *successive approximation* used for solving functional equations appears to be promising here as well. Let us start again from an arbitrary function V on the R.H.S. of the equation and see what function is obtained on the L.H.S. Denote our map by **M**: $V \to MV$, that is,

$$(7.10) \qquad \mathbf{M}V(x) = \max\{f(x, u) + \beta V(x^*) | \ u, \quad x^* = g(x, u)\}.$$

Under suitable compactness conditions, map **M** maps continuous functions into continuous functions. Introduce the space of continuous functions defined on a compact interval I of $\mathbf{R}^n$ and define the maximum norm on it: $\|V\| = \max_{x \in I} |V(x)|$. We allude at the end of Appendix A that this space is complete with respect to the maximum norm, it is a Banach space. If a sequence of continuous functions $\{V_k\}$ *uniformly converges* to V, (that is, the critical index is independent of x), then the norms of differences converge to zero: $\lim_{k \to \infty} \|V_k - V\| = 0$. We can repeat the procedure (7.10) *ad infinitum*. Starting from an arbitrary function V_0 we want to approximate V by $V_k = \mathbf{M}V_{k-1}$. In parallel, we solve the maximum problem (7.10) and denote the obtained feedback function as h_k.

Theorem 7.5.* *(Blackwell, 1965, Stokey and Lucas, 1989, Theorem 4.7.) Let V_0 be an arbitrary function $I \to \mathbf{R}$ and let $V_k = \mathbf{M}V_{k-1}$ be the iteration in (7.10). Under suitable compactness and convexity assumptions, the successive approximations converge: $V_k \to V$ and $h_k \to h$.*

Instead of proof. We recall that the contraction map theorem (Theorem 3.4) was proved for the n-dimensional space and the remark following it, mentioned that the theorem is valid for abstract Banach spaces as well. At the successive approximation of differential equations (Theorem 5.2) we have already alluded to this circumstance. Blackwell (1965) worked out a theorem just to prove Theorem 7.5. ∎

On the analogy of Theorem 7.3 we can present

Theorem 7.6. *Under suitable smoothness conditions, the following optimality conditions are necessary:*

$$(7.11) \qquad \frac{\partial f}{\partial u}[x, h(x)] + \beta \frac{\partial V}{\partial x}\{g[x, h(x)]\} \frac{\partial g}{\partial u}[x, h(x)] = 0.$$

Having determined the optimal feedback and transition functions, the following natural questions can be asked: (i) Does there exists any steady state, characterized as the fixed point of the optimal transition function: $x^\circ = \phi(x^\circ)$? (ii) How many steady states exist? (iii) Is there any locally or globally stable steady state?

Stokey and Lucas (1989, Chapter 6) study these difficult questions. We shall return to them in Section 8.3, after studying concrete examples in Section 8.1 and 8.2.

7.2 STOCHASTIC PRINCIPLE OF OPTIMALITY

Already at the deterministic problem, we have encountered a good deal of mathematical problems, which we neglected. The difficulties become even more pronounced at the stochastic problems. It is characteristic that Stokey and Lucas (1989) had to work out two long mathematical chapters (Chapters 7 and 8) before embarked on the mathematically correct discussion of their Chapter 9.

We shall only illustrate some of the difficulties. In the classical probability theory a random variable is either discrete or continuous. For example, the result of tossing is 0 (Head) or 1 (Tail), while the annual quantity of precipitation in Budapest can be characterized by a positive real number within the interval of several hundred millimeters. In practice we have only data in millimeters, that is, discrete values, but in theory this discretization is unacceptable. For example, the expected value of the tossing is $1/2$, moreover, the limit distribution of a series of discrete distributions is frequently continuous.

Even this duality between discrete and continuous quantities is disturbing as the duality of the whole book amply demonstrates. There is, however, a mixture of discrete and continuous distributions: for example, in the desert the annual quantity of precipitation is zero with positive probability and positive with a continuous distribution. Moreover, within the continuous distributions there are two types, namely, *absolute continuous* and *singular* distributions.

There is a further complication if we deal with time series of dependent random variables. After Theorem 1.7 we have already mentioned the simplest case, namely, that of homogeneous Markov chains with finite state space. The treatment of the general case of *Markov processes*, however, would go beyond the scope of the book.

For simplicity's sake, we shall only consider *series of time-independent random vectors*. Let z_t and w_t be the k-dimensional random variables of time-period t which modify the value of the objective function and the control variable, respectively. Assuming that z_t and w_t are known at the beginning of period t, relations (7.1)–(7.2) are replaced by

$$(7.12) \qquad \mathbf{E}\left\{ f_{T+1}(x_{T+1}, z_{T+1}) + \sum_{t=0}^{T} f_t(x_t, u_t, z_t) \right\} \to \max$$

subject to

$$(7.13) \quad x_{t+1} = g_t(x_t, u_t, w_t), \qquad t = 0, 1, \ldots, T \quad \text{and} \quad x_0 \quad \text{given}$$

where $\mathbf{E}$ stands for the *expectation operator*. Correspondingly, (7.4) is replaced by the generalized value function equation:

$$v_t(x_t) = \max\{ f_t(x_t, u_t, z_t) + \mathbf{E}v_{t+1}(x_{t+1}) |\ u_t,\ x_{t+1} = g_t(x_t, u_t, w_t) \}.$$

As illustrations, the stochastic generalizations of Example 7.1 and of Problem 7.1 are presented.

Example 7.4. Elementary stochastic example. $T = 1$, $n = 1$, state space equation: $x_1 = x_0 + u_0$ and $x_2 = x_1 - u_1 - w_1$; loss function: $\mathbf{E}(x_1^2 + x_2^2)$, $v_1(x_1) = \min\{\mathbf{E}x_2^2|\ u_1\} = \min\{\mathbf{E}(x_1 - u_1 - w_1)^2|\ u_1\} = \mathbf{E}(w_1)^2$ if $u_1 = x_1$. $v_0(x_0) = \min\{x_1^2 + \mathbf{E}w_1^2|\ u_0\} = \mathbf{E}w_1^2$ if $u_0 = -x_0$.

Problem 7.4. Elementary stochastic problem. $T = 1$, $n = 1$, state space equation: $x_1 = x_0 + u_0 + w_0$ and $x_2 = x_1 + u_1 + w_1$; loss function: $\mathbf{E}(x_0^2 + u_0^2 + x_1^2 + u_1^2 + x_2^2) \to$ minimum.

We present the stochastic generalization of Theorem 7.4:

Theorem 7.7. (Stokey and Lucas, 1989, Chapter 9). Assume that functions f and g satisfy concavity and compactness conditions. Then the stochastic optimal feedback function is time-invariant: $u = h(x, w)$ and, together with the value function V, satisfies the following functional equation:

$$V(x) = \max\{f(x, u, z) + \beta \mathbf{E}V(x^*, w)|\ u, \quad x^* = g(x, u, w)\}$$

Remarks. 1. If the perturbations vectors are not independent of each other, then the conditional expectations and regressions replace the absolute expectations in (7.12)–(7.13), respectively. For normal distributions, the conditional distributions are also normal, thus the conditional expectations are relatively easily calculated by recursions.

2.* The method of successive approximation can be extended from the deterministic to the stochastic case.

7.3. OPTIMAL LQ CONTROL WITH PERFECT OBSERVATION

In the subsequent Sections we shall consider the most important special case of general control theory: the system is *linear*, the objective function is *quadratic* and the distributions are *Gaussian*, for short, LQG system. Section 1.4 treated the controllability and observability of deterministic linear systems. These two concepts are independent of optimization but the latter is strictly based on the former.

Let $x \in \mathbf{R}^n$ be the *state* vector and $u \in \mathbf{R}^m$ *control* vector. For the time being, it is supposed that state x_t is precisely known at the beginning of time-period t. Let A_t and B_t be $n \times n$ and $n \times m$

matrix series, respectively, $t = 0, \ldots, T$. Our discrete-time, linear time-variant system is characterized by the following
State space equation

(7.14) $x_{t+1} = A_t x_t + B_t u_t,$ x_0 given.

We repeat the formula for time-additive utility/loss function:

(7.1) $$f_{T+1}(x_{T+1}) + \sum_{t=0}^{T} f_t(x_t, u_t).$$

The simplest loss function, which gives interior optimum, is the *quadratic* one:

(7.15) $f_t(x_t, u_t) = (x_t - x_t^*)^{\mathbf{T}} F_t(x_t - x_t^*) + (u_t - u_t^*)^{\mathbf{T}} G_t(u_t - u_t^*)$

where $\{x_t^*, u_t^*\}$ is the so-called *reference path*, F_t and G_t are n- and m-dimensional symmetric positive definite matrices (except for $G_{T+1} = 0$), their entries showing the losses of time-period t caused by unit deviations of state and control variables from their reference values.

The basic task in LQ is to minimize the time-additive quadratic loss function (7.1) [and (7.15)] under the constraint of the linear state space equation (7.14).

In contrast with physics where the sign of most quantities is arbitrary, in economics the bulk of the variables are positive. For example, in the latter it is not indifferent if output is lower or higher than the reference value. Therefore in reasonable economic applications the reference paths are chosen so that for all the parameter values the feasible paths remain either below or above the reference paths, this issue of sign does not matter. From now on it is assumed that the reference path is identically zero: $x^* = 0$ and $u^* = 0$.

Theorem 7.8. *(Kalman; Bryson and Ho, 1969, Section 2.3.) The optimal solution of the LQ problem is given by the following Optimal feedback*

$$u_t^\circ = -K_t x_t^\circ.$$

Optimal feedback matrix

(7.16) $K_t = (B_t^{\mathbf{T}} S_{t+1} B_t + G_t)^{-1} B_t^{\mathbf{T}} S_{t+1} A_t$

where

(7.17)
$$S_t = A_t^{\mathbf{T}} S_{t+1} A_t - K_t^{\mathbf{T}} (B_t^{\mathbf{T}} S_{t+1} B_t + G_t) K_t + F_t,$$
$$S_{T+1} = F_{T+1}.$$

The minimal remaining loss between t and $T+1$ is a quadratic function of the optimal state x_t°:

(7.18)
$$v_t(x_t^{\circ}) = x_t^{\circ\mathbf{T}} S_t x_t^{\circ}.$$

Remarks.   1. It can be proved by mathematical induction that the foregoing inverses exist. Indeed, S_{T+1} is a positive definite matrix and in iterations (7.16)–(7.17), this property is preserved. The matrix to be inverted, $B_t^{\mathbf{T}} S_{t+1} B_t + G_t$ is the sum of a nonnegative and a positive definite matrices, that is, their characteristic roots and the sum of the roots are positive, that is, the matrix is invertible.

2. Note that the feedback matrix of the optimal feedback can be determined before starting the actual control (*off-line*), and the recursion goes in reversed time, S_{t+1} yields K_t, and it yields S_t, and so on. This feature is quite common in dynamic optimization. This complex calculation, generally requiring computer, is independent of the initial state, which is unknown at the calculation. If the off-line calculations had been done before the real-time (*on-line*) control started, then the actual control can be determined as a function of the already known initial state.

3. We could have solved the problem directly with the help of quadratic programming, but then the solution may have required much more time. Our dynamic programming procedure makes use of the fact that considering the state space equation as a single large system (Theorem 7.1), the corresponding coefficient matrix is quasi-block-diagonal and that of the objective function is block-diagonal. If we were interested in only one solution, then the step-size would be $(nT)^3$. Utilizing the special structure, the iterative solution reduces the step-size to $n^3 T$.

4. The time-invariant *asymptotic problem* is an important special case: $T \to \infty$. Then the objective function needs normalization: $\lim_T (\sum_t f_t / T)$. For appropriate assumptions, K and S satisfy the timeless algebraic matrix equation:

(7.16*) $K = (B^{\mathbf{T}} S B + G)^{-1} B^{\mathbf{T}} S A,$

(7.17*) $S = A^{\mathbf{T}} S A - K^{\mathbf{T}} (G + B^{\mathbf{T}} S B) K + F.$

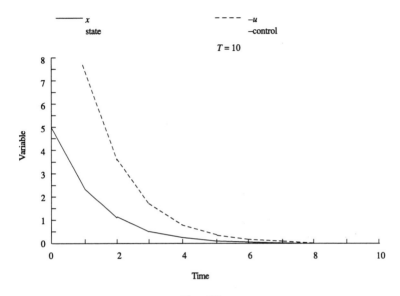

Figure 7.2
Long term optimal LQ control

It can be shown that then the feedback is stable, that is, $\rho(A - BK) <$ 1, and $\lim_t x_t = 0$ (Preston, 1977).

As a continuation of Figure 7.1, we present the asymptotic case (approximated by $T = 10$) of Example 7.2 in Figure 7.2.

Proof. Using the experience with linear-quadratic problems, we conjecture that the remaining loss is a quadratic function of the current state [(7.18)]. Substituting (7.14) and (7.15) into (7.5)–(7.6)

$$(7.19) \quad v_t = x_t^T F_t x_t + u_t^T G_t u_t + (A_t x_t + B_t u_t)^T S_{t+1} (A_t x_t + B_t u_t),$$

$$(A_t x_t^\circ + B_t u_t^\circ)^T S_{t+1} B_t + u_t^{\circ T} G_t = 0.$$

Rearranging the last equation:

$$(7.20) \qquad u_t^\circ = -(B_t^T S_{t+1} B_t + G_t)^{-1} B_t^T S_{t+1} A_t x_t^\circ.$$

The multiplier of x_t^o in (7.20) is just matrix $-K_t$ in (7.16). We have to confirm the conjecture in (7.18). Plugging (7.20) into (7.19), a quadratic form $x_t^{o\,\mathbf{T}} S_t x_t^o$ is obtained where (7.17) gives the recursion.∎

7.4* OPTIMAL LQG CONTROL WITH IMPERFECT OBSERVATION

This Section contains a verbal description of the stochastic generalization of optimal LQ control, denoted by LQG.

Optimal state estimation

The first step in this direction is to add an additive stochastic disturbance to the state space equation.

For the time being, control is switched off: $u_t = 0$. We also assume that the distributions of initial state and the disturbances are independent.

In addition to the stochastic disturbance affecting the state space equation, we assume that the observation of the state is also imperfect. Supplementing the deterministic imperfection described by (1.37), a stochastic disturbance appears.

The method of *least squares* is well known from the literature. In an optimal estimation, we have to minimize the various errors.

A fundamental theorem of LQG (Kalman filter, 1960; Bryson and Ho, 1969, Section 12.4) states that the optimal state estimation can be done in a sequence.

Like in optimal control, in the optimal state estimation a large part of the computations can be done off-line. Here the direction of the time is the usual.

Optimal stochastic control

Until now we have touched optimal control under full information and optimal state estimation in an imperfectly observed system with zero control. We shall now mention the unification of the previous two tasks: under imperfect observation, we optimize control *and* estimation. The task is to minimize the quadratic loss function under the constraint of the stochastic version of state space equation (7.14) and the stochastic observation equation.

According to another fundamental theorem (Kalman, 1960, Bryson and Ho, 1969, Section 14.7) the optimal solution of the double LQG problem is a sequence of estimations and a feedback control which is based on it, they are basically independent of each other.

It is of historical interest that in the solution of this problem, at least in an important special case, economists (Simon (1956) and Theil (1957)) overtook the engineers in proving the so-called certainty equivalence theorem.

Other control problems

The problems discussed above are only the simplest ones. In the remaining part of this Chapter we list several issues, with the corresponding sources.

a) It is much more difficult to study but at the same time much more realistic to assume *multiplicative* rather than additive disturbances (Wonham, 1967).

b) The classical control theory discussed so far assumes that the system is *perfectly centralized and has perfect memory*. Witsenhausen (1968) gave a counterexample showing that in a *nonclassical* optimum control problem the optimal control is generally nonlinear. The logic of the example is rather simple: in *decentralized* control the value of the control variable may serve as a *signal*. We give a simple practical example. Some years ago, even in our age of computers, it was not easy to learn who sent me what sum in the Hungarian banking system. A trivial solution would have been to ask client i, $i = 1, 2, 3$ to send the already rounded-off sum with a slight correction: to add i cents to the bill. This is a very cheap solution but it could be improved, at least in principle, if the signal were $.1i$ cents, $.01i$ cents, and so on.

c) Until now it was assumed that the distributions of the system's parameters are given. What happens if the statistical examinations start simultaneously with the control? This issue belongs to *dual* or *adaptive* control theory (for example, Tse and Athans, 1967).

d) The present Chapter was confined to LQG problems. What happens in the general case of nonlinear state space equation or nonquadratic loss function? This issue is taken up in the next Chapter.

8. APPLICATIONS OF DYNAMIC PROGRAMMING

The previous Chapter outlined the mathematical theory and the control-theoretic application of the dynamic programming, now some economic applications follow. Sections 8.1 and 8.2 discuss optimal saving (exogenous factor prices) and accumulation (endogenous factor prices), respectively. Section 8.3 extends the analysis of Section 8.2 to the n-dimensional case. Section 8.4 presents a game-theoretic application of dynamic programming (due to Levhari and Mirman, 1980). Detailed description is to be found in Sargent (1987), Manuelli and Sargent (1987) and Stokey and Lucas (1989) which is used again extensively.

8.1 OPTIMAL SAVING

First we shall investigate the problem of optimal saving. It will be seen that it is a reducible problem discussed before Theorem 7.1.

Let $u(c) : \mathbf{R} \to \mathbf{R}$ be a concave felicity function, β $(0 < \beta \leq 1)$ the discount factor and $U(c) = \sum_{t=0}^{T} \beta^t u(c_t)$ the utility of consumption path $\{c_t\}$. Let k_t be the initial capital at the beginning of time-period t, w_t the *wage*, c_t the *consumption* at time-period t and r the time-invariant *interest factor*. By definition, we have *the capital dynamics*

$$k_{t+1} = r(k_t + w_t - c_t).$$

Therefore the optimization problem can be rewritten as

$$\sum_{t=0}^{T} \beta^t u\left(k_t + w_t - \frac{k_{t+1}}{r}\right) \to \max$$

where

$$rk_t + rw_t \geq k_{t+1}, \qquad t = 0, 1, \ldots \quad \text{and} \quad k_0 \quad \text{given}, \qquad k_T = 0.$$

Again, we shall distinguish the finite and infinite horizon problems.

Finite horizon

We shall solve this problem without dynamic programming.

Theorem 8.1. *If $\Psi = \beta r < 1$ and there is consumer credit, then the optimal consumption is decreasing with age.*

Proof. Allowing negative accumulated capital, the problem can be reduced to a single constraint: the discounted present value of lifetime consumption is equal to the sum of the initial capital and the discounted present value of lifetime earnings. Using the Lagrange multiplier method where π is a positive constant, the Lagrange function is as follows:

$$L(c_0, \cdots, c_T, \pi) = \sum_{t=0}^{T} [\beta^t u(c_t) + \pi r^{-t}(w_t - c_t)] + \pi k_0.$$

The necessary condition for interior optimum is

(8.1) $$u'(c_t) = \pi \Psi^{-t}.$$

Since u' is a decreasing function, c_t is also a decreasing function of age ($c_{t+1} < c_t$) if and only if $\Psi < 1$. ∎

As an illustration, we present

Example 8.1. Cobb–Douglas utility function: $u(c) = \log c$ and consumer credit. From (8.1), $c_t = \Psi^t/\pi$. π can also be calculated from $k_0 + \sum_{t=0}^{T} r^{-t} w_t = \pi^{-1} \sum_{t=0}^{T} \beta^t$.

For numerical presentation, let $T = 15$, $w = 1$, $k_0 = 1$, $r = 1.03$ and $\beta = 0.95$. Figure 8.1 displays that the consumer uses credit between years 7 and 15.

If there is no credit, then the temporary balances cannot be eliminated: $k_t \geq 0$, $t = 0, 1, \ldots, T$. Then the initial capital and the earnings determine whether the previous optimal path hits the constraints or not.

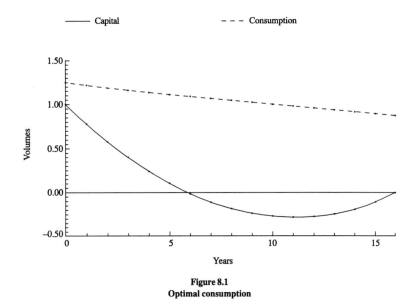

Figure 8.1
Optimal consumption

Before presenting the next problem, we shall introduce an important class of utility functions called Constant Relative Risk Aversion (for short, CRRA) utility functions (for more details, see Section 10.1 or C.2): $u(c) = \sigma^{-1}c^{\sigma}$ ($\sigma \neq 0$) where $\zeta = 1 - \sigma > 0$ is the coefficient of *relative risk aversion* and $\varepsilon = 1/\zeta$ is the elasticity of substitution. Realistically, $0 < \varepsilon < 1$. (Note that it is the coefficient σ^{-1} that makes $u(c)$ increasing for negative σ's.)

Problem 8.1. Solve Example 8.1 for CRRA utility function.

Uncertain life-time

Consider a person who has just retired and divide his uncertain life-span after retirement into two periods: $T = 1$. Assume that his per-period pension is equal to $b > 0$. Excluding bequest motive and consumer credit, we have $k_0 > 0$, $k_1 \geq 0$, $k_2 = 0$. Having formulated the constraints, we can turn to the objective function. Let q be the survival probability at the beginning of period 1: $0 < q < 1$. We shall work with an *expected undiscounted* CRRA utility function: $U(c_0, c_1) = \sigma^{-1}(c_0^{\sigma} + qc_1^{\sigma})$. We shall need notations $\alpha = k_0/b$ and $\Psi = (qr)^{\varepsilon}$.

Example 8.2. Pensioner with two-period retirement. a) If there is no capital depletion $(k_1 \geq 0)$, then the optimal consumption pair is

$$c_0 = \frac{r^2 k_0 + rb + b}{r + \Psi} \quad \text{and} \quad c_1 = c_0 \Psi.$$

b) There is no capital depletion if $\Psi(r\alpha + 1) \geq 1$. c) If $\Psi < 1$, then there is a critical ε_o such that for any $0 \leq \varepsilon \leq \varepsilon_o$, the condition holds.

Indeed, for the time being, we shall neglect the no-borrowing condition $k_1 \geq 0$. Then the two budget constraints can be unified into a single one: $r^2 k_0 + r(b - c_0) + b - c_1 = 0$. Noticing that q plays the role of the discount factor β, Problem 8.1 yields the relations in a). Borrowing is superfluous if $k_1 \geq 0$, or equivalently, if $c_1 \geq b$. By substitution, condition $q^\varepsilon (r^2 k_0 + rb) \geq r^{1-\varepsilon} b$ is obtained, yielding b).

Note that for capital depletion, the case of decreasing consumption is interesting: $\Psi < 1$. For $\varepsilon = 0$, the condition in c) reduces to $r\alpha \geq 0$.

Returning to the ratio $c_1/c_0 = \Psi$, note that this is a decreasing function of the elasticity ε. The smaller elasticity, the more balanced consumption path is.

It is the irony of fate that there are economists (see, for example Gokhale et al., 1996), who see this problem the other way around.

Problem 8.2. If there are neither annuities, nor intertemporal substitution, then our pensioner uses up the initial capital in two 'equal' parts, thus in both periods the consumption is equal to $c = (r^2 k_0)/(r + 1)$ and the expected value of the *unintended* bequests is $qk_1 = qrk_0/(r+1) > 0$. Calculate the potential annuity b and compare it to c.

Introduction of annuitized pensions leads to increased consumption and zero unintended bequest. This is an important piece in Gokhale et al.'s explanation for the chronic undersaving in the present US. Although I disagree with this explanation, furthermore, I consider the annuatization is a good rather than a bad thing, the present model is capable to tell such a story.

Infinite horizon

What happens if $T \to \infty$?

As a warm-up, we present

Example 8.3.   Cobb–Douglas utility function and constant earnings. Because of infinite horizon, Example 8.1 simplifies. For $r > 1$, $w(1 - 1/r)$ is finite and $\pi^{-1} = k_0 + w(1 - 1/r)(1 - \beta)$.

Considering general utility functions and excluding consumer credit, Stokey and Lucas proved

Theorem 8.2. *(Stokey and Lucas, 1989, Section 5.17.) For infinite horizon, general utility function and constant earnings, the consumption is decreasing with age, and depending on the initial capital, sooner or later is restricted to the current income.*

The next two problems use linear utility function (criticized in Section 10.3 below), just to highlight the significance of credit rationing.

Problem 8.3. Linear utility function with credit (Stokey and Lucas, 1989, p. 74). Let the initial capital be k_0 and the per-period utility function be $u(c) = c$. Assume that the consumer can save and borrow without limits at interest factor $r = 1/\beta$. Prove that the value function can be ∞ as well as $V(k) = k$.

Problem 8.4. Linear utility function without credit (Stokey and Lucas, 1989, p. 76). We modify Problem 8.3 at a single point, there is no consumer credit. Prove that the value function is $V(k) = k$ but the permanent no-consumption strategy does not satisfy the transversality condition.

8.2 ONE-SECTOR OPTIMAL ACCUMULATION

The model

Having touched the problem of exogeneous factor prices, we turn to the issue of endogeneous ones. Let T be the horizon of the model, finite or infinite. Let $u(c)$ be the felicity function, β the discount factor and $U(c) = \sum_{t=0}^{T} \beta^t u(c_t)$ the utility of consumption path c. The production function is denoted by $f^*(k)$ and it is assumed that in a time-period t, δk_t capital evaporates, therefore $c_t = f^*(k_t) + (1 - \delta)k_t - k_{t+1}$. Introducing the extended production function $f(k) = (1 - \delta)k + f^*(k)$, $c_t = f(k_t) - k_{t+1}$. (It could also be said that the old capital evaporates in a time-period, that is, $\delta = 1$ and $f^* = f$.) Our problem has the following structure:

(8.2) $$\sum_{t=0}^{T} \beta^t u[f(k_t) - k_{t+1}] \to \max$$

where

(8.3) $0 \le k_{t+1} \le f(k_t)$, $t = 0, 1, \ldots$ and k_0 given.

It is assumed that both the utility function and the production function are smooth, increasing and concave, and they satisfy the Inada-conditions:

$$U'(0) = \infty, \qquad U'(\infty) = 0, \qquad f'(0) = \infty, \qquad f'(\infty) = 0.$$

Furthermore, $0 \le \beta \le 1$.

It is useful to make a dictionary between the general and the specific terms. State: $x \to k$; control: $u \to c$; derived reward function: $\psi \to u$.

The general problem (8.2)–(8.3) has no explicit solution, however, the qualitative features can be analyzed and the model can be solved numerically. The following special case has a nice analytical solution (see the corresponding Problem 6.2 in continuous time).

For Cobb–Douglas utility function and production function: $u(c) = \log c$ and $f(k) = k^\alpha$, (8.2)–(8.3) reduce to

(8.4) $$\sum_{t=0}^{T} \beta^t \log(k_t^\alpha - k_{t+1}) \to \max$$

subject to

(8.5) $$0 \le k_{t+1} \le k_t^\alpha.$$

Finite horizon

Returning to the general framework (8.2)–(8.3), first we shall investigate the case of finite horizon: $T < \infty$. Then one can use the ordinary technique of maximization. In our case, due to the concavity of f and u, the necessary conditions are sufficient.

(8.6) $\beta f'(k_t) u'[f(k_t) - k_{t+1}] = u'[f(k_{t-1}) - k_t]$;
(8.7) $k_{T+1} = 0$, k_0 given.

We shall break up the issue into Examples and Problems.

Example 8.4. (Stokey and Lucas, 1989, pp. 11–2.) For the Cobb–Douglas problem, (8.6) reduces to

$$(8.8) \qquad \frac{\beta \alpha k_t^{\alpha-1}}{k_t^\alpha - k_{t+1}} = \frac{1}{k_{t-1}^\alpha - k_t}.$$

Introducing variable $z_t = k_t/k_{t-1}^\alpha$ and parameter $\Psi = \alpha\beta$, the solution can be guessed:

$$(8.9) \qquad z_t = \Psi \frac{1 - \Psi^{T-t+1}}{1 - \Psi^{T-t+2}}$$

and proved by mathematical induction. Plugging back yields the equation of the optimal capital accumulation:

$$(8.10) \qquad k_{t+1} = \Psi \frac{1 - \Psi^{T-t}}{1 - \Psi^{T-t+1}} k_t^\alpha.$$

Problem 8.5. a) Demonstrate that the second-order difference equation (8.8) can be reduced to a first-order equation (8.10) with substitution $z_{t+1} = 1 + \Psi - \Psi/z_t$. b) How can one find out (8.10)?

Problem 8.6. Show that (8.10) is a solution of (8.9).

Infinite horizon

We turn now to the infinite horizon problem. With problem (8.4)–(8.5) we can illustrate the various methods of solution of dynamic programming problems: (i) the extension of finite horizon, (ii) the determination of the value function with guessing or (iii) iteration.

General theorems suggest that the solution of the infinite horizon problem is the limit of the solutions of the corresponding finite horizon problems. This is illustrated by

Example 8.5. For $T = \infty$, the solution of Example 8.4 is given by

$$(8.11) \qquad k_{t+1} = \lim_{T\to\infty} \Psi \frac{1 - \Psi^{T-t}}{1 - \Psi^{T-t+1}} k_t^\alpha = \Psi k_t^\alpha$$

and the transversality condition also holds.

It is instructive to calculate the fixed point: $k^\circ = \Psi^{1/(1-\alpha)}$.

Returning to the general case, we shall use the functional equation (7.9).

$$(8.12) \quad V(k) = \max\{u[f(k) - k^*] + \beta V(k^*) | \qquad 0 \le k^* \le f(k)\}.$$

We have to solve this equation for the unknown value function. This is done in

Theorem 8.3. *(Stokey and Lucas, 1989, Sections 2.1 and 5.1.) Under appropriate assumptions on the production function and the utility function, we have the following*
Optimum-condition

(8.13) $$u'[f(k) - h(k)] = \beta V'[h(k)],$$

Envelope-condition

(8.14) $$V'(k) = f'(k)u'[f(k) - h(k)].$$

Equation (8.13) "equates the marginal utility of consuming current output to the marginal utility of allocating it to capital and enjoying augmented consumption next period". Equation (8.14) "states that the marginal value of current capital, in terms of total discounted utility, is given by the marginal utility of using capital in current production and allocating its return to current consumption" (Stokey and Lucas, 1989, p. 14).

We shall now present the guessing method.

Example 8.6. Guessing the value function (Stokey and Lucas, 1989, pp. 11–3). It can be checked that (8.11) is indeed a solution to the special version of (8.13)–(8.14). Indeed, (8.11) implies $u(c_t) = \log[(1 - \Psi)k_t^\alpha] = \log(1 - \Psi) + \alpha \log k_t$. Taking the logarithm of (8.11): $\log k_t = \log \Psi + \alpha \log k_{t-1}$. By (1.16), the solution is $\log k_t = \text{const} + \alpha^t \log k_0$, that is, $V(k_0) = \sum_t \beta^t u(c_t) = \text{const} + \alpha(1 - \Psi)^{-1} \log k_0$, $h(k) = \Psi k^\alpha$. Taking the derivative: $V'(k) = \alpha(1 - \Psi)^{-1}k^{-1}$, and so on.

Problem 8.7. Determine the constant term in $\log k_t$.

We shall now illustrate the method of successive approximation.

Example 8.7.* (Stokey and Lucas, 1989, pp. 93–5). One can prove that

$$\log k_t \leq \alpha^t \log k_0.$$

Take the transformation

$$\mathbf{M}V(k) = \max\{\log(k_t^\alpha - k^*) + \beta V(k^*)| \quad 0 \leq k^* \leq k_t^\alpha\}$$

and demonstrate with direct calculation that for $V_0 = 0$,

$$\mathbf{M}^t V_0(k) = \frac{1 - \beta^{t+1}}{1 - \beta} \log(1 - \Psi) + \frac{\Psi \log \Psi}{1 - \Psi} + \frac{\alpha \log k}{1 - \Psi}.$$

Asymptotically

$$V(k) = \frac{1}{1 - \beta} \log(1 - \Psi) + \frac{\Psi \log \Psi}{1 - \Psi} + \frac{\alpha \log k}{1 - \Psi}$$

which is just the solution.

According to Theorem 7.4, the transition function $\phi(k) = \Psi k^\alpha$ of the time-invariant saving ratio generates the optimal capital accumulation path.

Stochastic production function

Let $\{z_t\}$ be a series of identically and independently distributed random variables. It is assumed that at the beginning of time-period t, capital stock k_t and multiplicative disturbance z_t yields output $z_t f(k_t)$. We have the following stochastic dynamic programming problem:

$$\mathbf{E} \sum_{t=0}^{\infty} \beta^t u[z_t f(k_t) - k_{t+1}] \to \max$$

where

$$0 \le k_{t+1} \le z_t f(k_t), \qquad t = 0, 1, \dots \quad \text{and} \quad k_0 \quad \text{given.}$$

The deterministic value function equation (8.12) is replaced by its stochastic counterpart:

$$(8.15) \quad V(k, z) = \max\{u[z f(k) - k^*] + \beta \mathbf{E} V(k^*)| \qquad 0 \le k^* \le z f(k)\}.$$

Studying (8.15) leads us to the optimal capital stock $k^* = \phi(k, z)$. For interior optimum and differentiability, (8.15) is replaced by

$$V'[z f(k) - \phi(k, z)] = \beta \mathbf{E} V'[\phi(k, z)].$$

What is the solution for the special case?

Example 8.8. Stochastic production, Cobb–Douglas case and infinite horizon (Stokey and Lucas, 1989, Section 2.2). It can be shown that the optimal transition is given by the generalization of formula (8.11):

$$k_{t+1} = \Psi z_t k_t^\alpha.$$

Taking logarithms again:

$$\log k_{t+1} = \log \Psi + \log k_t^\alpha + \log z_t.$$

Generalizing the proof of Problem 8.7 yields

$$\log k_t = \log \Psi \frac{1 - \alpha^t}{1 - \alpha} + \alpha^t \log k_t + \sum_{i=1}^{t} \alpha^{i-1} \log z_{t-i}.$$

If z_t is, for example, a lognormal random variable with parameters $(0, \sigma)$, then the distribution of $\log k_t$ can also be determined. A nice result is obtained asymptotically: the limit of the deterministic term is equal to $(1 - \alpha)^{-1} \log \Psi$. The distribution of the stochastic term converges to a distribution. It is important that the influence of the initial deterministic case evaporates, the system is stochastically stable, *ergodic*.

8.3* MULTISECTOR OPTIMAL ACCUMULATION

This Section extends the investigation of Section 8.2 from a one-sector model to the multisector model of optimal accumulation, with paying a particular attention to the connection of the discount factor and the complexity of the optimal path.

Let us reformulate the one-sector growth model as follows: $k^* = f(k) - c$, that is, if $\mathcal{K}$ is the interval of feasible initial stocks, then the set of feasible terminal stocks k^* is $T_k = \{f(k) - c | \; c > 0\}$.

Jumping to the n-sector model, let $\mathcal{K}$ be an n-dimensional nonempty set of $\mathbf{R}^n$, which is the feasible set of stock vectors. Set T is a nonempty closed and convex subset of $\mathcal{K} \times \mathcal{K}$. For any vector $k \in \mathcal{K}$, consider $T_k = \{k^* \in \mathcal{K} | \; (k, k^*) \in T\}$ which is assumed to be nonempty and compact. Let $u : T \to \mathbf{R}$ be a bounded, continuous and strictly concave function, representing the *utility* of moving from k to k^*. The discount factor again lies between 0 and 1. The mathematical form of the problem is as follows:

$$\sum_{t=0}^{\infty} \beta^t u(k_t, k_{t+1}) \to \max$$

where

$$(k_t, k_{t+1}) \in T.$$

Let us start the review with the turnpike results of the 1960's. We shall need the notion of golden rule path, already mentioned in Section 6.1. The golden rule path is a stationary path maximizing the value of consumption $c = f(k) - k^*$. The first result concerns arbitrary number of sectors and weak discounting, while the second refers to a one-sector model with arbitrary discounting. Building on Samuelson (1965) and Cass (1966) and other studies, we have

Theorem 8.4. *(McKenzie, 1986.) For a given technology T and derived utility function u, if the discount factor β is close enough to 1, then the optimal path converges to the golden rule path.*

Theorem 8.5. *(Dechert, 1984.) For a one-sector optimal growth model, the convergence prevails at any discount factor.*

The following question logically arises: what happens to the multisector growth model if the discount factor is not close enough to 1? The researchers obtained a surprising result, namely, that in sufficiently general optimal growth models *everything goes*. We mention without proof a pioneering result.

Theorem 8.6. *(Boldrin and Montrucchio, 1986.) Consider an arbitrary function with continuous second derivative. Then there exists a growth model including a discount factor, such that the resulting optimal transition function is equal to the given function.*

Remark. There is a striking similarity between this statement and the static proposition of Debreu, Mantel and Sonneschein (see Section 6.3), severely undermining the central role which individual optimization plays in macroeconomics.

Theorem 8.6 is illustrated by

Example 8.9. (Boldrin and Montrucchio, 1986.) For $\phi(k) = 4k(1 - k)$, there exists a model with $\beta = 0.01$, such that the optimal transition function is equal to $\phi(k)$. As is known, this implies chaotic dynamics.

We encounter the following question. Why should we use such a strong discounting? Two answers can be given: a) the method is weak or b) there is no complicated dynamics for realistic discount factors. Sorger (1992) gives the second answer.

Theorem 8.7. *(Sorger, 1992, Theorem 1.)* Let $\{k_t\}_{t=0}^{\infty} \in \mathbf{R}^n$ *be an optimal growth path. Assume that there is a scalar solution* $z_1, \ldots, z_t, \ldots$ *to the following* $(n+1)$-*dimensional system:*

$$(8.16) \qquad \sum_{t=1}^{\infty} z_t \beta^{*t} = 1,$$

$$(8.17) \qquad \sum_{t=1}^{\infty} z_t \beta^{*t} k_t = k_0,$$

$$z_1 \geq z_2 \cdots \geq z_t \cdots \geq 0.$$

Then $\beta \leq \beta^*$.

Sketch of the proof. The starting step is Bellman-equation (7.9):
$$V(k) = \max\{u(k, k^*) + \beta V(k^*)| \ (k, k^*) \in \mathcal{T}\}$$
where the maximum at the R.H.S. is attained at a unique stock vector $k^* = \phi(k)$. ∎

How can one use Theorem 8.7? For simplicity's sake, let us consider the one-dimensional case: $n = 1$ when both systems are scalar:

$$(8.16^*) \qquad \sum_{t=1}^{\infty} z_t \beta^{*t} = 1,$$

$$(8.17^*) \qquad \sum_{t=1}^{\infty} z_t \beta^{*t} k_t = k_0.$$

Try with the simplest case when all z_t's are the same. From (8.16^*) $z_t \equiv (1 - \beta^*)/\beta^*$. Inserting into (8.17^*), yields

$$(8.18) \qquad \sum_{t=1}^{\infty} \beta^{*t}(k_t - k_0) = 0.$$

Corollary. *(Sorger, 1992, Corollary 1.)* Let ϕ map interval $I = [0, 1]$ into itself and let ξ be its greatest maximum. If $k^* = \phi(k)$ is a transition function of an optimal growth model, then the discount factor is at most $\beta^* = 1 - \xi$.

Proof. Plugging $k_0 = \xi$, $k_1 = 1$ and $k_2 = k_3 = \cdots = 0$ in (8.18), $\beta^* = 1 - \xi$ is obtained. ∎

As an illustration, consider a chaotic map (of Section 3.3).

Example 8.10. (Sorger, 1992.) Consider the most famous logistic function as a transition function: $\phi(k) = 4k(1-k)$, then $\xi = 1/2$, that is, $\beta^* = 1/2$. Therefore the chaotic logistic function cannot be an optimal transition function if $\beta > 1/2$. On the other hand, according to Deneckere and Pelikan (1986), there is a chaotic model for any $\beta \leq 1/4$.

Problem 8.8. (Sorger, 1992, pp. 168–9.) Prove that the chaotic logistic function cannot be an optimal transition function if $\beta \geq 0.475$ which is quite modest, at least on an annual basis. (Hint: Choose the initial value $k_0 = [1 + \cos(3\pi/8)]/2$.)

Sorger (1992) remarks that his results are far from being sharp. There is a growing literature on the problem (the latest piece is Mitra and Sorger, 1999). We shall pick up the most interesting result which underlines the exact connection between turbulence (introduced in Section 3.3) and the discount factor.

Theorem 8.8.* (Mitra, 1998.) Let $(\mathcal{T}, u, \beta)$ be a dynamic optimal growth model with transition function ϕ. Map ϕ generates turbulence if and only if $\beta \leq 1/4$.

8.4 THE GREAT FISH WAR

This Section presents a most interesting application of dynamic programming to game theory, following Levhari and Mirman (1980). The starting point is the *cod war* between Iceland and the U.K. The prize in this war was the quantity and the distribution of catch between the two countries. To give a feeling on the order of magnitude, in the 1950's the stock and the annual catch of the fish equaled 2.5 million and 250 thousands tons, respectively. In the 1990's these numbers dropped to 600 thousands and 150 thousands tons, respectively.

The model

Let k_t be the stock at the beginning of time-period t. Suppose that in the absence of fishing, the stock would grow according to the biological law

(8.19) $$k_{t+1} = k_t^\alpha \quad \text{where} \quad 0 < \alpha < 1.$$

Obviously, the biological steady state is equal to $k^\# = 1$, which is a matter of normalization.

There are, however, fishers around, moreover, they are citizens of two competing nations, indexed by $i = 1, 2$. Let us denote $c_{i,t}$ the catch of country i at t. Then (8.19) is modified to

$$(8.20) \qquad k_{t+1} = (k_t - c_{1,t} - c_{2,t})^\alpha.$$

We assume that country i has a Cobb–Douglas utility function with discount factor β_i, horizon T:

$$U_i = \sum_{t=0}^{T} \beta_i^t \log c_{i,t}, \qquad i = 1, 2.$$

We also assume that each country takes the strategy of the other country as given, that is, both play Cournot–Nash strategies. Formally: If country i takes country j's catch $\{c_{j,t}\}$ as given, then it can maximize its utility function U_i under the constraint (8.20). If there exist two series $\{c_{1,t}\}$ and $\{c_{2,t}\}$ such that each is optimal with respect to the other, then they are called a *Cournot–Nash optimum*.

Solution

Before presenting the optimum for horizon T, we shall introduce the following notations:

$$R_T = \sum_{t=0}^{T} \alpha^t, \quad \psi_i = \alpha\beta_i, \quad S_{i,T} = \sum_{t=0}^{T} \psi_i^t.$$

Theorem 8.9. *(Levhari and Mirman, 1980.) For time-horizon T, the optimal Cournot–Nash policy of time-period t is given by*

$$(8.21) \qquad c_{i,t}^{\circ} = \frac{\psi_j S_{i,T-1-t}}{S_{1,T-t} S_{2,T-t}} k_t^{\circ}, \qquad i = 1, 2, \quad j \neq i.$$

Proof.   By mathematical induction with respect to T.   ∎

Our result is simplified if we return to infinite horizon.

Theorem 8.10. *(Levhari and Mirman, 1980.) For infinite time-horizon, the optimal Cournot–Nash policy is independent of time:*

$$(8.22) \qquad c_{i,t}^o = \frac{\psi_j(1 - \psi_i)}{\psi_1 + \psi_2 - \psi_1\psi_2} k_t^o, \qquad i = 1, 2, \quad j \neq i,$$

while the steady state stock is given by

$$(8.23) \qquad k^o = \left(\frac{1}{\psi_1} + \frac{1}{\psi_2} - 1\right)^{\alpha/(\alpha-1)}$$

Proof. The asymptotic value of (8.21) is (8.22). Then the stock at the initial of time-period t is easily calculated:

$$k_t = \left[\frac{\alpha\beta_1\beta_2}{\beta_1 + \beta_2 - \alpha\beta_1\beta_2}\right]^{R_t} k_0^{\alpha t}, \qquad i = 1, 2.$$

Asymptotically, (8.23) is obtained. ∎

Remark. Observe that the weaker is the discounting (that is, the larger the discount factor), the larger is the fish stock.

Problem 8.9. Check the calculations.

Problem 8.10. a) Solve the problem if the two countries cooperate and having a joint discount factor $(\beta_1 = \beta_2 = \beta)$, divide the catch equally. b) Demonstrate that in this case for a given fish stock, they catch less than in the competitive case, but the steady state stock is so much higher than under competition $(k^* > k^o)$ that the steady state catch is still higher $(c^* > c^o)$.

Example 8.11. Numerical illustration. Let $\alpha = 0.5$ and $\beta = 0.95$. Then $k^o = 0.29$; $c^o = 0.103$ and $k^* = 0.45$; $c^* = 0.124$. The catch ratios (c/k) at the two steady states (realized with or without cooperation) are 0.355 and 0.275, respectively. To illustrate the dynamics, let $k_0 = 0.5k^o$ be the initial state in terms of the competing steady state. The paths of competition and cooperation are illustrated in Figure 8.2. It is remarkable that at the initial stage, the catch under competition is larger than that under cooperation, but the drop in the total stock reverses the process.

In addition to the symmetrical Cournot–Nash solution, duopoly theory also investigates the asymmetric Stackelberg equilibrium. Here it is the leader which takes the first step, and then the follower makes the second step. In this game the leader has a better position than the follower. Instead of going into details, we refer to a more important issue.

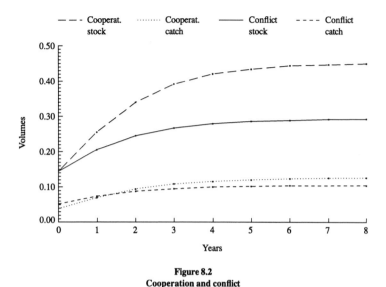

Figure 8.2
Cooperation and conflict

Time inconsistency

Consider a dynamical system with a control vector consisting of two blocks, because the system is controlled by two decision makers. At the Cournot–Nash solution the principle of optimum remains valid, but at the Stackelberg solution the validity disappears: Even for $s \leq t$, the follower's decisions $u_{1,s}$ depend on the leader's decisions $u_{2,t}$. This is called *time-* or *dynamic inconsistency.*

Rather than going into details, we present a simple example from life (Kydland and Prescott, 1977). Suppose that in time-period 0 the authorities forbid that people build houses in a flood area, because the cost of building dams would be much higher than the saving due to utilizing substandard area. People knowing time-inconsistency, nevertheless would build their houses in the flood area in time-period 1. They calculate that the authorities realize that after the construction of these houses, the total value of the menaced houses will be much higher than the cost of the dams, thus the authorities would rush to build the dams. However, the authorities may stick to their decisions, and let the illegally built houses be washed away by the flood, to ensure long-run credibility.

Or to give a military example, Ms Thatcher behaved in a time-consistent way during the Falkland war of 1982 when she spent a sum much exceeding the value of the Falkland islands to liberate the territory from the Argentine army. In this way she upheld the credibility of Britain (and other countries) to punish the aggressor.

Other problems

Chapter 5 of Stokey and Lucas (1989) contains a lot of economic problems solvable by dynamic programming. We only list few problems. Cake eating. Growth with technological development. Cutting tree. Learning by doing. Accumulation of human capita. Growth with human capital. Investment with convex costs. Investment with constant returns to scales. Optimal growth with recursive preferences. The (s, S) inventory policy.

Because of space limitations, it is much more difficult to give economic applications to LQG control which can be summarized in our book. We call the reader's attention to the following sources: Athans (1972), (1975), Pitchford and Turnovsky (1977), Salmon and Young (1979) in Holly et al. (1979), Chow (1981) and Kendrick (1981).

9. OPTIMAL CONTROL

After discussing discrete-time optimal systems in Chapters 7 and 8, we turn to continuous-time optimal systems. We are given a control system (see Section 5.4), that is, a system of differential equations, describing the dynamics of the state vector as a function of a control vector. For a given control path, the state space equation determines the state path. In the theory of *Optimal Control*, the equations are time-variant and we are given a scalar-valued function, called *reward function* of time, of the state vector and of the control vector. We are looking for that control path which maximizes the time integral of the reward function. Section 9.1 outlines the solution of the basic problem. Section 9.2 presents an important special case of optimal control, namely, the *Calculus of Variations*. Section 9.3 contains additional material, including sufficient conditions, the isoperimetric problem and the present value problem. We shall rely on Kamien and Schwartz (1981) (see also Chiang, 1992). For a more demanding treatment, see Pontryagin et al. (1961) and Gelfand and Fomin (1962).

9.1 BASIC PROBLEM

Let $\mathbf{T} = [0, T]$ be a real interval, figuratively we shall speak of time-interval. Let $f(t, x, u)$ be a smooth real valued function defined on the domain $\mathbf{T} \times \mathbf{R}^{n+m}$ where x and u are n- and m-dimensional vectors of state and control variables, respectively. Let us consider the pair of smooth vector functions $\{x, u\}$ defined on interval $\mathbf{T}$ (figuratively, paths) where the initial state $x(0) = x_0$ is given and the end state

$x(T) = x_T$ is free. Paths $\{x(t), u(t)\}$ satisfying these conditions are called *feasible paths* where $[x(t), u(t)]$ lies in the *feasibility set* $\mathcal{X} \subseteq \mathbf{R}^{n+m}$ which has a nonempty interior.

The *state space equation* is as follows: let $g : \mathbf{R} \to R^{1+n+m}$

(9.1) $\dot{x}(t) = g[t, x(t), u(t)]$ where x_0 is given.

Let the *objective function* be

(9.2) $I[x, u] = \displaystyle\int_0^T f[t, x(t), u(t)]\, dt.$

This is called a *functional*, because it is defined on a function, namely on $\{x, u\}$ rather than on points, like $(x(t), u(t))$.

As is known from the ordinary calculus of $\mathbf{R}^n \to \mathbf{R}$ functions, there are two types of optima: local and global, and it is much easier to determine the former than the latter. This is also the case with our functional. Mainly, we are looking for the *local* maximum satisfying (9.1) and the optimal path $\{x, u\}$ belongs to the interior of the feasibility set.

Through the state space equation, the control path determines the state path, notation: $x[t, u(t)]$, therefore the functional can be simplified:

(9.3) $I[u] = \displaystyle\int_0^T f[t, x(t, u(t)), u(t)]\, dt.$

The following theorem determines the necessary condition of the local optimum.

Theorem 9.1. *(Pontryagin, et al., 1961.) If functional I has a local optimum at the feasible function u, then the triple of functions $[x, u, p] : \mathbf{R} \to \mathbf{R}^{n+m+n}$ have to satisfy the following (differential) equations on interval $\mathbf{T}$:*
a) the state space equation

(9.1) $\dot{x}(t) = g[t, x(t), u(t)]$ where x_0 is given,

b) the multiplier equation

(9.4) $\dot{p}(t)^{\mathbf{T}} = -f_x[t, x(t), u(t)] - p(t)^{\mathbf{T}} g_x[t, x(t), u(t)]$

with the end condition $p(T) = 0$;

c) *the optimality condition*

(9.5) $f_u[t, x(t), u(t)] + p(t)^T g_u[t, x(t), u(t)] = 0.$

Remark. It is to be underlined that the state space equation (9.1) as well as the multiplier equation (9.4) is an ordinary differential equation. Moreover, the latter has an end condition rather than an initial condition, thus it should be solved in reversed time, from T to 0. Such a reversal in time is very frequent in dynamic optimization problems, see Chapters 7 and 8.

It is easier to remember the results if the *Hamiltonian function* or simply, *Hamiltonian* is introduced:

(9.6) $H[t, x, u, p] = f(t, x, u) + p^T g(t, x, u).$

Then we have the Hamiltonian form of
a) the state space equation

(9.1H) $\dot{x} = H_p^T \quad \text{with} \quad x(0) = x_0;$

b) the multiplier equation

(9.4H) $\dot{p} = -H_x^T \quad \text{with} \quad p(T) = 0;$

c) the optimality condition

(9.5H) $H_u = 0.$

Outline of proof. We shall give two proofs. The idea of the first proof is due to Euler (see also the proofs of Theorems 5.1 and 7.1), which is simple but imprecise. The second is due to Lagrange, which is sophisticated but precise.

The discrete approximation of Euler

To simplify notation we shall confine ourselves to the scalar case, but the procedure is easily extended to the general case of (n, m). Divide interval **T** into k equal subintervals where the length of each subinterval is equal to $h = T/k$. Replace our continuous-time functions and equations with their discrete counterparts, as we did at the proof of Theorem 5.1. (Here we do not denote the dependence on k.) Let $t_i = ih$, $x_i = x(t_i)$, $u_i = u(t_i)$, $f_i(x_i, u_i) = f(t_i, x_i, u_i)$ and $g_i(x_i, u_i) = g(t_i, x_i, u_i)$. Then the discretized equations are

(9.1D) $$x_{i+1} = x_i + hg_i(x_i, u_i), \qquad i = 0, \ldots, k - 1,$$

(9.2D) $$I = h \sum_{i=0}^{k-1} f_i(x_i, u_i) \to \max.$$

Let $x = (x_0, \ldots, x_{k-1})^{\mathbf{T}}$, $u = (u_0, \ldots, u_{k-1})^{\mathbf{T}}$, $p = (p_0, \ldots, p_{k-1})^{\mathbf{T}}$ be k-vectors. Using the method of Lagrange multipliers, we arrive at the following unconditional maximum problem:

$$L(x, u, p) = \sum_{i=0}^{k-1} \Big\{ h f_i(x_i, u_i) - p_i[x_{i+1} - x_i - hg_i(x_i, u_i)] \Big\}.$$

After introducing $H_i(x_i, u_i, p_i) = f_i(x_i, u_i) + p_i g_i(x_i, u_i)$, take the partial derivatives of L with respect to p_i, x_i and u_i: in addition to (9.1D) we have

$$h \frac{\partial H_i}{\partial x_i} + p_i - p_{i-1} = 0, \qquad i = 0, \ldots, k - 1,$$

$$h \frac{\partial H_i}{\partial u_i} = 0, \qquad i = 0, \ldots, k - 1.$$

Divide all equations by h and let k go to ∞. Returning to our continuous-time functions, we obtain the Hamiltonian form (9.1H), (9.4H) and (9.5H). There is only a single gap in the proof: it is not proved that the approximation is correct, that is, the optimum exists and it is the limit of the optima of the discrete approximations.

Variational method of Lagrange

We shall follow again the Lagrange method but with a twist.

$$\int_0^T f[t, x(t, u(t)), u(t)] \, dt = - \int_0^T p(t)^{\mathbf{T}} \dot{x}(t) \, dt$$
$$+ \int_0^T H[t, x(t, u(t)), u(t)] \, dt.$$

We integrate by parts the first term at the R.H.S. and plug the result into $I[u]$:

(9.7)
$$I[u] = \int_0^T H[t, x(t, u(t)), u(t)]\, dt$$
$$+ \int_0^T \dot{p}(t)^{\mathbf{T}} x(t)\, dt - p(T)^{\mathbf{T}} x(T) + p(0)^{\mathbf{T}} x(0).$$

At this point we shall apply the *variational method*, which is a congenial generalization of the usual maximization procedure. Let u° be the feasible local maximum (time-function) and let u^* be another feasible time-function. Let $v = u^* - u^\circ$ be the deviation from the optimum. Let a be a scalar in $[0,1]$ and let $u^\circ + av$ be an intermediate control path. Denote $y(t, a)$ the corresponding state path. Finally, let

(9.8)
$$G(a) = I[u^\circ + av]$$

be a scalar function. According to the usual condition of a local maximum with respect to a, $G(a)$ has a local maximum at $a = 0$, hence $G'(0) = 0$. Assuming that the functions are differentiable (it needs a proof) and using the derivation rules of the parametric integral and of the composite function, from (9.7) we arrive at

$$G'(0) = \int_0^T \left[(H_x + \dot{p}^{\mathbf{T}}) y_a + H_u v \right] dt - p(T)^{\mathbf{T}} y_a(T) = 0.$$

Let us choose $p(t)$ such a way that the expressions before y_a and $y_a(T)$ disappear, that is, to have the multiplier equation (9.4H) with the end condition $p(T) = 0$. Hence condition $G'(0) = 0$ simplifies into

$$\int_0^T H_u v\, dt = 0.$$

We want to prove that $H_u \equiv 0$, [(9.5H)]. Assume the contrary, for example, $H_u > 0$ on a subinterval $J = (\delta, \varepsilon)$. Since v is an arbitrary time-function with $v(0) = v(T) = 0$, let $v(t) = (t - \delta)(\varepsilon - t)$ on J and 0 otherwise. Then a contradiction is obtained:

$$\int_0^T H_u v\, dt > 0. \qquad \blacksquare$$

Until now we had a fixed initial condition and a free end condition. What happens if the end state is fixed or the end time is free? Equation (9.7) in the second proof suggests the answer which is called *boundary and transversality conditions*. a) If both the end time T and the end state x_T are fixed, then $p(T) = 0$ is replaced by $x(T) = x_T$ in (9.4). b) If the end time T is free but the end state $x(T) = x_T$ is fixed, then $p(T) = 0$ is replaced by $H(T) = 0$ in (9.4). (For details, see Kamien and Schwartz, 1981, II.7.)

In the following we shall show simplest examples on the application of Theorem 9.1.

Example 9.1. Scalar linear-quadratic problem. $f(t, x, u) = x^2 + u^2$, $\dot{x} = u$, $x(0) = 1$, $x(1) = 1$. From (9.6): a) $\dot{x} = u$, b) $\dot{p} = -2x$, c) $p = -2u$. Take the derivative in a) and c): $\ddot{x} = \dot{u}$, $\dot{p} = -2\dot{u}$. Using b): $2x = \dot{p} = -2\ddot{x} \Rightarrow \ddot{x} = x$. The characteristic roots are as follows: $\lambda_{1,2} = \pm 1$, thus the solution is $x(t) = \xi_1 e^t + \xi_2 e^{-t}$. From the initial conditions, $x(0) = \xi_1 + \xi_2$ and $x(1) = \xi_1 e + \xi_2 e^{-1}$, that is, $\xi_1 = 1/(e+1)$, $\xi_2 = e/(e+1)$.

The following problem points out an interesting property of the Hamiltonian.

Problem 9.1. Show that for an autonomous problem where both f and g are time-invariant, the Hamilton function is constant along the optimal path.

Functions satisfying the necessary conditions of Theorem 9.1 are called *extremal paths* without the implication that they are true maxima or minima.

At the end of the Section we discuss the economic meaning of the multiplier which is very similar to that encountered in static problems: *shadow price*. Let $V(0, x_0)$ be the maximum value of the problem with initial condition $(0, x_0)$. Using the tricks in the proof above, for smooth functions, it can be shown that $p(0) = V_x^T(0, x_0)$. In words: the initial value of the multiplier is equal to the derivative of the objective function with respect to the initial state. Since our objective function is time-additive, the identity holds at any instant $t \in \mathbf{T}$ and for any feasible state x, $p(t) = V_x^T(t, x)$.

9.2 CALCULUS OF VARIATIONS

Historical introduction

The theory of optimal control was created by Pontryagin et al. around 1960, but an important special case, the *Calculus of Variations* was already born before 1700. Moreover, certain elementary problems in the calculus of variations were already solved in the ancient times.

The theory of optimal control simplifies into the calculus of variations if (i) the dimensions of the state and control vectors are the same: $n = m$ and (ii) the state space equation is the simplest: $g(t, x, u) = u$, that is, $\dot{x} = u$. Then the reward function is equal to $f(t, x, \dot{x})$. (Problems of this type were already mentioned in Section 7.1 for discrete-time systems.)

It occurs rather often that the independent variable is the horizontal coordinate rather than the time. Then x is traditionally replaced by y and t by x. Nevertheless, we shall always stick to the (t, x) plane.

There is no doubt that the simplest problem in the calculus of variations is the following: What is the shortest path between two points? *The straight line connecting the two points.* In the language of geometrical optics: In a homogeneous space the light takes the shortest path.

A similar problem is the reflection of an object in a mirror. Here the optimal path consists of two straight lines with equal angles of incidence and reflection (Figure 9.1).

Snellius in 1620 and Descartes in 1637 analyzed refraction. They discovered that at the point of refraction, the ratio of sinuses of the incidence and the refraction angles is equal to the ratio of the velocities of light in the water to that in the air (Figure 9.2).

These problems were generalized by Fermat around 1650: How does the light goes from point A to point B in an inhomogeneous space? *In the fastest path.* This is called *Fermat's principle*.

A couple of economic problems can be solved by the methods of calculus of variations (Chapter 10), thus we shall outline it. As a warm-up, however, we shall solve three elementary problems mentioned above.

Problem 9.2. a) Using elementary geometry, prove that the straight line is the shortest path between two points. b) Try the analytical method of Theorem 9.1.

9. Optimal Control

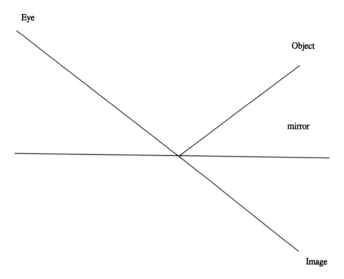

Figure 9.1
Mirror and minimum time

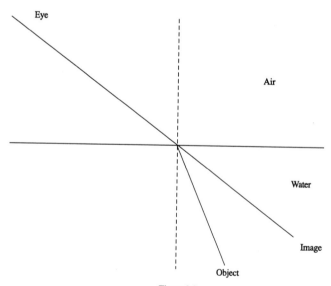

Figure 9.2
Refraction of light and minimum time

Problem 9.3. Consider a planar mirror. Using elementary geometry, prove that between point O(bject) and its mirror I(mage), the light chooses the shortest path (the incidence angle is equal to the reflection angle).

Problem 9.4. Refraction. Let an Object be under the water level and the observer's Eye be over the water level. Then the observer will see the Image above the Object. Using elementary calculus, prove that at refraction the light chooses the fastest path.

Simonyi (1981, pp. 192–4) illustrates the difficulties of that age by a rather sophisticated but elementary contemporary geometrical proof of Problem 9.4. Pólya (1968, Chapter IX) excels with an elegant proof based on a mechanical analogy. In these problems the solution consists of one or two straight lines, therefore elementary proofs are possible. In general, however, the solution is a genuine curve, thus elementary proofs do not help.

The basic problem

We turn to the basic problem of the calculus of variations. Let x be an $\mathbf{R} \to \mathbf{R}^n$ function and let

$$(9.9) \qquad I[x] = \int_0^T f[t, x(t), \dot{x}(t)] \, dt$$

be a functional to be maximized. The initial point $x(0) = x_0$ and the end point $x(T) = x_T$ are given.

Theorem 9.2. (Euler–Lagrange, 1744–1755.) If functional I in (9.9) has a maximum at a feasible function x, then function x satisfies the following, Euler–Lagrange differential equation:

$$(9.10) \qquad f_x(t, x, \dot{x}) = \frac{d}{dt} f_{\dot{x}}(t, x, \dot{x})$$

where f_x and $f_{\dot{x}}$ are the partial derivatives of f.

Outline of the proof. According to Theorem 9.1, the multiplier equation (9.4) yields $\dot{p}^{\mathbf{T}} = -f_x$ and the optimum condition (9.5) yields $f_{\dot{x}} + p^{\mathbf{T}} = 0$. Take the time derivative of the second and plug into the first:

$$0 = \frac{d}{dt}(f_{\dot{x}} + p^{\mathbf{T}}) = \frac{d}{dt} f_{\dot{x}} - f_x. \qquad \blacksquare$$

Remarks. 1. Executing the derivations in (9.10), the following second-order vector differential equation is obtained:

$$f_x(t, x, \dot{x}) - f_{t\dot{x}}(t, x, \dot{x}) - f_{x\dot{x}}(t, x, \dot{x})\dot{x} - f_{\dot{x}\dot{x}}(t, x, \dot{x})\ddot{x} = 0.$$

2. In contrast to the ordinary optimization, it is quite difficult to prove an existence theorem in the calculus of variations. For example, the following simple problem has no solution: Find the *longest* path connecting two points.

Many textbooks on calculus of variations solve various problems with the help of (9.10). We shall not go into the details of these problems except the oldest one:

Example 9.2.* Brachistochron (Johann Bernoulli, 1696). Given a vertical plane representing a homogeneous gravitational field, and two points in that plane and the connecting straight line is neither horizontal, nor vertical. Determine that connecting planar curve which assures the fastest connection between the upper and lower points.

Around 1630 Galilei gave a wrong solution, claiming that the optimal curve is an arc in circle (Simonyi, 1981, p. 172). As Pólya (1968, Chapter IX) outlines, Johann Bernoulli, generalizing Fermat's principle of Problem 9.4 in a congenial way, gave a correct solution in 1696. Foreshadowing Euler's discrete approximation, Bernoulli divided the vertical distance between the two points into k equal parts and assumed that the velocity is constant within each zone. Using the solution of Problem 9.4 $(k - 1)$-times, he obtained an optimal broken line. In the limit, he obtained the cycloid:

$$t = \delta + b(v - \sin v) \quad \text{and} \quad x = \alpha + b(1 - \cos v), \quad 0 \le v \le 2\pi, \quad b > 0.$$

Boyer (1968) discusses the quarrel of Jacob and Johann Bernoulli (the brothers) on priority. At that time there were very few scientific journals, thus it was customary to pose problems in letters. The Bernoullis sent their problem, among others, to Newton. When Newton returned by post the correct solution, the scientific world was relieved that the aging giant recovered from his long illness.

9.3 SUPPLEMENTS

In this Section we present some additional material.

Sufficient conditions

Similarly to the ordinary optimization, in optimal control there are also second-order conditions (Kamien and Schwartz, 1981, pp. 122–3).

Theorem 9.3. *If the reward function f and the governing function g are concave in x as well as in u, and $p(t) \geq 0$, then the necessary condition is also sufficient.*

Remark. Arrow proved that it is sufficient if for a given p, the *maximized Hamilton function* $H^\circ(t, x, p) = \max_u H(t, x, u, p)$ is concave in x. (Kamien and Schwartz, 1981, 204–5).

Sketch of the proof. Let $^\circ$ be the distinguishing notation for the optimum and drop the time symbol. Since f is a concave function of (x, u),

$$f^\circ - f \geq f_x^\circ(x^\circ - x) + f_u^\circ(u^\circ - u),$$

that is,

$$\int_0^T (f^\circ - f)\, dt \geq \int_0^T \left[f_x^\circ(x^\circ - x) + f_u^\circ(u^\circ - u) \right] dt.$$

Using the multiplier equation (9.4) with $p \geq 0$ and the optimum condition (9.5) at the R.H.S., we can integrate by parts the terms containing $\dot{p}$. Taking into account the state space equation (9.1), the R.H.S. is equal to

$$\int_0^T p^{\mathbf{T}} \left[g^\circ - g - g_x^\circ(x^\circ - x) + g_u^\circ(u^\circ - u) \right] dt,$$

and this is nonnegative by the concavity of g, that is,

$$\int_0^T (f^\circ - f)\, dt \geq 0. \qquad \blacksquare$$

Deficient reward functions

Both in optimal control and calculus of variations the process of solution is much easier if the reward function is *deficient*, that is, if one of the three variables t, x, $u = \dot{x}$ does not appear in f. Confining our attention to the calculus of variations, we study all the three cases:

(i) The reward function does not depend on t: $f(x, \dot{x})$. According to the Euler–Lagrange differential equation (9.10), there exists a constant c, such that the extremal solution satisfies the implicit differential equation $f(x, \dot{x}) - \dot{x} f_{\dot{x}}(x, \dot{x}) = c$ as can be verified directly from Remark 1 to Theorem 9.2.

(ii) The reward function does not depend on x: $f(t, \dot{x})$. Again by (9.10), identity $df_{\dot{x}}(t, \dot{x})/dt = 0$ holds, that is, there exists a constant c yielding the implicit differential equation $f_{\dot{x}}(t, \dot{x}) = c$.

(iii) The reward function does not depend on $\dot{x}$: $f(t, x)$. Again by (9.10), $f_x(t, x) = 0$ holds, maximizing the reward function at each instant. Of course, such an optimum only exists if the side conditions are located on the curve obtained.

Until now we have only discussed unconstrained problems. We now discuss several other problems.

Isoperimetric problem

Let us consider a more complex part of calculus of variations. Let $f : \mathbf{R}^{2n+1} \to \mathbf{R}$ be a reward function and $g : \mathbf{R}^{2n+1} \to \mathbf{R}$ be a constraint function. The *generalized isoperimetric problem* consists of the following: functional

$$(9.9) \qquad I[x] = \int_0^T f[t, x(t), \dot{x}(t)] \, dt$$

is to be maximized under the integral condition

$$(9.11) \qquad J[x] = \int_0^T g[t, x(t), \dot{x}(t)] \, dt = \kappa$$

where κ is a scalar.

The solution is given by the usual method of Lagrange-multiplier:

Theorem 9.4. *The generalized isoperimetric problem.* If functional (9.9) is maximized under the functional constraint (9.11), and the solution is not an unconstrained extremum of (9.11), then there exists a real number p, such that function x satisfies the unconstrained Lagrange problem $L = f + pg$, that is, it satisfies the Euler–Lagrange differential equation concerning L.

Outline of the proof. Because of constraint (9.11), the constrained maximum of (9.9) is equal to the unconstrained maximum of the following problem in the calculus of variations:

$$I^*[x] = \int_0^T \Big\{ f[t, x(t), \dot{x}(t)] + pg[t, x(t), \dot{x}(t)] \Big\} \, dt - p\kappa.$$

The only assumption is that x° is not a solution to (9.11). ∎

Let us start the illustrations with an elementary but very difficult problem.

Problem 9.5.* The original isoperimetric problem (Pólya, 1968, Chapter X). Prove by an elementary method that among planar domains with a given perimeter, the circle has the maximal area (Figure 9.3).

Example 9.3. (Kamien and Schwartz, 1981, Problem I.7.2.) Solve the following variant of Problem 9.5*. We are given an interval **T** of length T and a real number $\kappa > T$. Find a curve $\{x(t)\}$ of length κ with an area above the interval is maximal.

Let $(0, 0)$ and $(0, T)$ be the two end points of the curve and let $(t, x(t))$ be an arbitrary point of the curve. Then we have the reward function $f[t, x(t), \dot{x}(t)] = x(t)$ and the integrand of the constraint, $g[t, x(t), \dot{x}(t)] = \sqrt{1 + \dot{x}(t)^2}$. (Note that by the Pythagoras theorem, the distance between $(t, x(t))$ and $(t + dt, x(t + dt))$ is given by $\sqrt{1 + \dot{x}(t)^2} \, dt$.)

We have a Langrange function $L = x - p\sqrt{1 + \dot{x}^2}$. The corresponding relation (9.10) gives

$$1 = -\frac{d}{dt} \frac{p\dot{x}}{\sqrt{1 + \dot{x}^2}}.$$

Following the method of solution of a separable differential equation (Theorem 5.4), we obtain (with a constant k)

$$t = -\frac{p\dot{x}}{\sqrt{1 + \dot{x}^2}} + k.$$

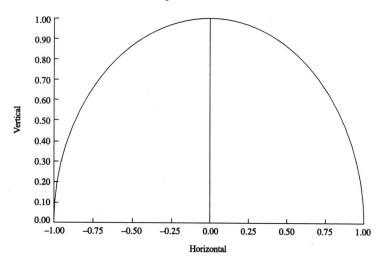

Figure 9.3
Izoperimetric problem

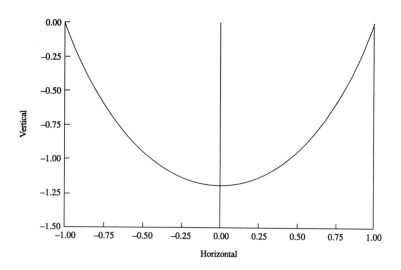

Figure 9.4
Chain curve

Solving the algebraic equation for $\dot{x}$:

$$\dot{x} = -\frac{t-k}{\sqrt{p^2 - (t-k)^2}}.$$

Let $v = p^2 - (t-k)^2$, $dv = -2(t-k)dt$. Then (with a constant c)

$$x(t) = \int_0^T \dot{x}(t)\,dt = -\int v^{-1/2}\,dv/2 + c = -\sqrt{v} + c,$$

that is,

$$(x-c)^2 + (t-k)^2 = p^2,$$

which is an equation for the arc in circle. The completion of the proof is left to the reader.

The following problem occurs in mechanics.

Problem 9.6.* Chain curve, catenary (at the end of the 17th century). We have a homogeneous chain of length κ with end points $(-a, D)$ and (a, D), $\kappa > 2a$, $D > 0$. A mechanical principle suggests that the center of mass of the chain curve is at the minimal height. a) Prove that the reward and the constraint functions are $f[x(t), \dot{x}(t)] = x\sqrt{1 + \dot{x}(t)^2}$ and $g[x(t), \dot{x}(t)] = \sqrt{1 + \dot{x}(t)^2}$, respectively. b) Prove that the equation of the chain curve is $x(t) = A\cosh(\delta t) + \gamma$ where A, δ and γ are appropriate constants. (Reminder from calculus: $\sinh t = (e^{-t} - e^t)/2$ and $\cosh t = (e^{-t} + e^t)/2$.) c) What is the ordinate of the deepest point? (Figure 9.4 displays the chain curve for $a = 1$, $\kappa = 3$.)

It is interesting that it was again Galilei who initiated the problem and gave a wrong solution as in Example 9.2*: parabola (Boyer, 1968, p. 358).

The next problem is almost trivial but will have an economic application in Chapter 10.

Example 9.4. Degenerate isoperimetric problem. Consider the case when both f and g are independent of $\dot{x}$: $f(t, x)$ and $g(t, x)$. We have to solve the parametric equation

$$f_x(t, x) + pg_x(t, x) = 0$$

for $x(t, p)$ and the obtained function is to be substituted into the constraint (9.11):

(9.11°) $$J[x] = \int_0^T g[t, x(t, p)]\,dt = \kappa.$$

If (9.11°) has a solution for p, notation: $p°$, then the single-variable function $x(t, p°)$ is the solution to the original problem.

Closed domain

Pontryagin and his associates have analyzed the most general problem, notably, they allowed that the control vector belong to a closed
domain $\mathcal{U}$ and the control function u be only piecewise continuous.
They proved the validity of Pontryagin's *maximum-principle* in this
framework which replaces the optimum-condition.

$$H[t, x^\circ(t), u, p(t)] \leq H[t, x^\circ(t), u^\circ, p(t)], \quad u \in \mathcal{U}$$

and the multiplier equation does not apply at the discontinuities.
There is another complication, namely, that in the Hamiltonian the
reward function is also multiplied by a nonnegative rather than a positive multiplier, namely, by $p_0 \geq 0$. To prove existence, in addition
to the usual smoothness and concavity condition it is assumed that
$g(t, x, u)$ is linear in u.

The following example is the simplest case when the optimum lies
on the boundary of the closed domain and it is a piecewise continuous
function. (Theorem 9.1 does not apply here.)

Example 9.5. Special time-minimum problem. Assume that a
mass point should be transferred from (0,0) into (0,1) in such a way
that the absolute value of the acceleration is at most 1, while the
initial and end velocities are 0. The time-minimum policy can be
formulated rather simply: until halfway maximal acceleration, from
halfway maximal deceleration.

More generally,

Example 9.6. General time-minimum problem. In certain military problems it is very important that an enemy missile be destroyed
in minimum time, while the control vector lies in an m-dimensional
domain. Then T is a free variable, $f(t, x, u) = 1$, that is, $I[u] = T - 0$.
Then an interior optimum satisfies
b) the multiplier equation:

$$(9.4') \qquad \dot{p}^{\mathbf{T}}(t) = -p(t)^{\mathbf{T}} g_x[t, x(t), u(t)], \quad x(T) = 0;$$

c) the optimum condition:

$$(9.5') \qquad p(t)^{\mathbf{T}} g_u[t, x(t), u(t)] = 0.$$

Further examination is needed to determine if there is an interior
optimum or not.

Present-Value Problems

It is quite frequent in economic problems that the rewards and sacrifices of different dates are *discounted* to a single date (Kamien and Schwartz, 1981, pp. 151–8).

$$(9.1^*) \qquad I[x, u] = \int_0^T e^{-\beta t} f(x, u) \, dt$$

where $\beta > 0$ is the *discount rate* and $g(x, u)$ is also independent of t. In this case the discounted Hamiltonian is

$$(9.6^*) \qquad H = e^{-\beta t} f(x, u) + p^T g(x, u),$$

and the conditions corresponding to (9.4H)–(9.5H) are as follows:

$$H_u = e^{-\beta t} f_u + p^T g_u = 0 \quad \text{and} \quad \dot{p}^T = -H_x = -e^{-\beta t} f_x - p^T g_x.$$

Then it is convenient to calculate at the *current value* rather than the discounted value. Introducing the *current multiplier*

$$\mathbf{p}(t) = e^{\beta t} p(t),$$

and the *current Hamiltonian*

$$\mathbf{H} = e^{\beta t} H = f(x, u) + \mathbf{p}^T g(x, u),$$

the following conditions are derived:

$$\mathbf{H}_u = 0 \quad \text{and} \quad \dot{\mathbf{p}}^T = \beta \mathbf{p}^T - \mathbf{H}_x.$$

Returning to the original formulas:

$$\dot{x} = g(x, u), \qquad f_u(x, u) + \mathbf{p}^T g_u(x, u) = 0,$$
$$\dot{\mathbf{p}}^T = \beta \mathbf{p}^T - f_x(x, u) - \mathbf{p}^T g_x(x, u).$$

Hence the advantage of the transformation can be seen: we arrive at an autonomous (time-invariant) differential equation.

At discounting problems it is quite usual that the horizon is infinite: $T = \infty$. Then difficulties arise at the transversality conditions (see Chiang, 1992) as we have seen at the discrete-time framework in Section 8.2.

10. OPTIMAL CONSUMPTION PATHS

One can solve various problems concerning continuous-time *optimal consumption paths* with the help of the theory of optimal control. Section 10.1 studies the optimal consumption path under given labor and capital incomes (Ramsey, 1928). Two alternative assumptions are considered: (i) certain and (ii) uncertain lifespan. Sections 10.2 and 10.3 analyze optimal consumption paths with endogenous incomes derived from a given production function, for infinite and finite horizons, respectively. We shall rely on Kamien and Schwartz (1981) and Simonovits (1995b) (Koopmans (1967) is also recomemnded).

10.1 EXOGENOUS WAGE AND INTEREST

This Section divides into two parts: certain and uncertain lifespans.

CERTAIN LIFESPAN

Let T be a positive real, and let us assume that the consumer lives T years and T is already known at the consumer's birth (Kamien and Schwartz, 1981, pp. 25-7). At time t denote his exogenous labor income $w(t)$, his capital $k(t)$, which yields capital income $rk(t)$ where r is the exogenous instantaneous interest rate. Denoting the current consumption by $c(t)$, we have the *current budget constraint*

$$(10.1) \qquad c(t) + \dot{k}(t) = rk(t) + w(t).$$

Utility functions

Let $u : \mathbf{R} \to \mathbf{R}$ be the *instantaneous utility function* (or equivalently, the so-called felicity function) and $u[c(t)]$ be the instantaneous utility of consumption $c(t)$. We shall assume that the utility of the lifetime consumption is the integral of the discounted instantaneous utility function $e^{-\beta t}u[c(t)]$ over $[0,T]$ where $\beta \geq 0$ is the *discount rate*. In formula:

$$(10.2) \qquad I[c] = \int_0^T e^{-\beta t}u[c(t)]\, dt.$$

We shall frequently use the *absolute and relative risk aversion coefficients* introduced at the expected utility function of Neumann-Morgenstern (Pratt in 1964, Arrow in 1965, see Arrow (1970) and Varian (1992, pp. 177–201)):

$$(10.3) \qquad a = \frac{-u''}{u'} \quad \text{and} \quad \zeta = ac.$$

Perhaps it will be helpful to refer to the role of risk in the terminology. The relative risk aversion coefficient shows how much the consumer is ready to pay with respect to his wealth to avoid a small variation in the utility of his consumption. In our case this quantity is equal to the amount by which the consumer is ready to diminish his lifetime consumption to avoid variations in consumption. Because of the concavity of the utility function, a and ζ are positive reals.

Example 10.1. Utility functions with constant absolute (CARA) and relative (CRRA) risk aversions.

$(10.4) \quad u(c) = a^{-1}e^{-ac} : \qquad$ CARA,

$(10.5a) \quad u(c) = \sigma^{-1}c^{\sigma} \quad \text{if} \quad \sigma \neq 0 : \qquad$ CRRA,

$(10.5b) \quad u(c) = \log c, \qquad \text{if} \quad \sigma = 0 : \qquad$ Cobb $-$ Douglas.

For a CRRA utility function, $\zeta = 1 - \sigma$.

Optimal consumption path

Inserting the balance equation (10.1) into $I[c]$, yields the regular problem in the calculus of variations of Section 9.2:

$$(10.6) \qquad I[k] = \int_0^T e^{-\beta t}u[rk(t) + w(t) - \dot{k}(t)]\, dt,$$

with the following side conditions:

(10.7) $k(0) = k_0$ and $k(T) = k_T.$

We are given both the initial and the closing stocks. Alternatively, the closing stock can be freed and then an additional function is to be added to (10.6).

We shall also need the following notations.
The integral of $e^{\xi t}$ on $[0, T]$:

(10.8) $J(\xi) = \dfrac{e^{\xi T} - 1}{\xi}$ if $\xi \neq 0,$ $J(0) = T;$

present value of lifetime earnings:

(10.9) $W = \displaystyle\int_0^T e^{-rt} w(t)\, dt,$

differential interest rate: $\delta = r - \beta.$

We shall assume that the lifetime earning is large enough to finance both the desired capital accumulation and some consumption:

(10.10) $W > e^{-rT} k_T - k_0.$

(10.10) trivially holds if $r = 0$, $k_T < k_0$ and $w(t) > 0$.
We announce the basic theorem of life-cycle theory.

Theorem 10.1. a) *For certain lifespan, the (relative) growth rate of the optimal consumption is equal to the ratio of the differential interest rate and the relative risk aversion coefficient:*

(10.11) $\dfrac{\dot{c}}{c} = \dfrac{\delta}{\zeta}.$

b) *For a CRRA utility function, the initial value of the optimal consumption path is given by*

(10.12) $c_0 = \dfrac{k_0 - e^{-rT} k_T + W}{J(\delta/\zeta - r)}.$

Remark.   In the modern economic literature first Modigliani and Brumberg (1954) discussed a similar problem in the framework of the theory of *life-cycle*, assuming discrete time, zero interest rate and no-discounting.

Proof.   a) Using the Euler–Lagrange differential equation [(9.10)] of the problem, we arrive to the following relation:

$$(10.13) \qquad \frac{d}{dt}\left[-e^{-\beta t}u'(c)\right] = e^{-\beta t}u'(c)r.$$

Execute the derivation of the product on the L.H.S: $\beta e^{-\beta t}u'(c)$ $- e^{-\beta t}u''(c)\dot{c}$. Applying notations (10.3) and δ, we obtain the optimality condition (10.11). In general, ζ depends on c, therefore the obtained differential equation cannot be solved in a closed form.

b) Using the assumption of CRRA [(10.5)], we can integrate the differential equation $\dot{c}/c = \delta/\zeta$: $c(t) = c_0 e^{\delta t/\zeta}$.

Reinserting into the balance equation: $\dot{k} - rk = w - c_0 e^{\delta t/\zeta}$.

Using now the method of multipliers mentioned in Section 5.4, multiply both sides of the last equation by e^{-rt} to obtain the derivative of a function on the L.H.S.:

$$\frac{d}{dt}\left[e^{-rt}k\right] = e^{-rt}(\dot{k} - rk) = we^{-rt} - c_0 e^{(\delta/\zeta - r)t}.$$

Integrating both sides of the new equation over $[0, T]$, and using notations W and J: $e^{-rT}k_T - k_0 = W - c_0 J(\delta/\zeta - r)$, implying (10.12).∎

Before closing this topic, the theorem is illustrated on an Example and a Problem.

Example 10.2. For Cobb–Douglas utility function ($\zeta = 1$), zero earnings ($w = 0$) and no bequest ($k_T = 0$), relation $\delta/\zeta - r = -\beta$ holds, that is

$$(10.12') \qquad c_0 = \frac{k_0}{J(-\beta)}.$$

Problem 10.1. Consider the following scenario: the expected remaining life expectancy at retirement $T = 20$ years, the unit annual pension $w = 1$ and initial retirement wealth $k_0 = 5$, zero closing capital: $k_T = 0$, neither discounting, nor interest: $\beta = r = 0$. Determine the optimal initial consumption for a CRRA utility function.

UNCERTAIN LIFESPAN

Certainly Yaari (1965) was the first who extended the theory of Modigliani and Brumberg (1954) from certain lifespan to uncertain lifespan. Following him, assume that the lifespan is a random variable, denoted as t, its distribution but not its value is known at birth. Let T be the maximal lifespan, and $Q(t)$ be the conditional probability of survival at age t. Thus function $Q(t)$ is nonincreasing, $Q(0) = 1$ and $Q(T) = 0$. For simplicity, we assume that the end stock of capital is zero: $k_T = 0$. We shall only consider the case when full life annuity is available which has no extra insurance cost.

Using the concept of expected present value, the problem can be reduced to the degenerate isoperimetric problem studied in Example 9.4. Indeed, the expected present value of the total lifetime consumption is equal to the expected present value of the total lifetime earning plus the initial capital:

$$(10.14) \qquad \int_0^T e^{-rt} Q(t) c(t)\, dt = \int_0^T e^{-rt} Q(t) w(t)\, dt + k_0.$$

We shall generalize (10.9) and (10.8) to stochastic lifespan.

$$(10.15) \qquad \mathbf{E}W = \int_0^T e^{-rt} Q(t) w(t)\, dt,$$

$$(10.16) \qquad \mathbf{E}J(\xi) = \int_0^T e^{-\xi t} Q(t)\, dt.$$

On the basis of the above relations, one can prove

Theorem 10.2. a) *For uncertain lifespan, zero bequest and full life annuity, the optimal consumption path is determined by the following equation:*

$$(10.17) \qquad c(t) = u'^{-1}(pe^{-\delta t})$$

where u'^{-1} is the inverse of the marginal utility function u' and the parameter value p can be determined from (10.14).

b) *For a CRRA utility function, the growth rate of optimal consumption path is equal to the corresponding value of certainty, and its initial value is a stochastic generalization of the deterministic case:*

$$(10.18) \qquad c_0 = \frac{k_0 + \mathbf{E}W}{\mathbf{E}J(\delta/\zeta - r)}.$$

Remark. A comparison of Theorems 10.1 and 10.2 reveals that the longevity risk (connected to the uncertainty of lifespan) does not really influence the optimal consumption path.

Proof.   a) Let

$$(10.19) \quad f(t,c) = e^{-\beta t}Q(t)u[c(t)], \quad g(t,c) = e^{-rt}Q(t)[w(t) - c(t)],$$
$$\kappa = k_0.$$

Relying on Example 9.4 and plugging (10.19) the optimum condition yields (10.14), via equation $e^{-\beta t}u'[c(t)] = pe^{-rt}$ we obtain

$$(10.20) \quad u'[c(t)] = pe^{-\delta t},$$

which, after inverting u', yields $c(t)$.

If we can solve equation (10.17) for $c(t)$, then via constraint (10.14), p and $c(t,p)$ are obtained.

b) This is the case under CRRA. Substitute $u'(c) = c^{-\zeta}$ into (10.20): $c(t)^\zeta = pe^{\delta t}$, that is,

$$(10.21) \quad c(t) = p^{1/\zeta}e^{\delta t/\zeta}.$$

Let $c_0 = p^{1/\zeta}$. Inserting (10.21) into (10.14) and applying notations (10.15)–(10.16), yields (10.18). ∎

Although the introduction of pension systems has dramatically increased the role of annuitized income, there still remains a large part of nonannuitized income. Yaari (1965) and others claimed that even in this case the optimum is interior, that is, savings never disappear before death.

Using the elaborate technique of optimal control with inequalities on the state variables, Leung (1994) has shown that the optimal consumption policy excludes interior optimum. While I accept Leung's mathematical analysis, I question the suitability of one economic assumption underlying his as well as other similar investigations. In my opinion, the main economic problem is not so much the depletion of wealth but the relatively low level of pension, implying low late consumption with respect to high early consumption. In my opinion, under a certain level of annuity, it is important to annuatize as much capital as possible. If the scope of annuatization is limited, then one has to choose a rather low risk strategy, avoiding both wealth depletion and unbalanced consumption (see Example 8.2).

10.2 ENDOGENOUS WAGE AND INTEREST: INFINITE HORIZON

In the previous Section both labor and capital incomes were given. From now on, we shall explain the dynamics of labor and capital incomes by a well-behaved production function.

Centralized solution

First it is assumed the "central planner" solves the aggregate optimization problem (Kamien and Schwartz, 1981, pp. 98–103). We shall now consider the case of infinite horizon: $T = \infty$. For simplicity's sake, we shall start from the per capita production function and assume away the death risk. Let k, c and f be the per capita capital, consumption, and output, respectively. Here $f(k)$ is the *per capita production function*, derived from the traditional production function $F(K, L)$. Under the usual assumptions, $f' > 0$ and $f'' < 0$. The new balance equation is $\dot{k} = f(k) - c$, which leads to the new problem in calculus of variations:

$$(10.22) \qquad I[k] = \int_0^\infty e^{-\beta t} u[f(k(t)) - \dot{k}(t)]\, dt.$$

Considering the related Euler–Lagrange differential equation (9.10), the new optimum condition is characterized by

Theorem 10.3. *For endogenous labor and capital incomes, the optimal consumption path is determined by*

$$(10.23) \qquad \frac{\dot{c}}{c} = \frac{f'(k) - \beta}{\zeta(k)} \quad \text{and} \quad \dot{k} = f(k) - c$$

where $\zeta(k)$ is the coefficient of relative risk aversion at $c = f(k) - \dot{k}$.

Remark.   1.   The transversality condition at the infinity is given by

$$\lim_{T \to \infty} k_T u'[c(T)] e^{-\beta T} = 0.$$

2. According to the marginal productivity theory, $r = f'(k)$, that is, (10.23a) corresponds to (10.11).

Proof.   First we search for the steady state (k°, c°). At this point, conditions $\dot{k} = 0$ and $\dot{c} = 0$ hold. According to the previous conditions and optimum condition (9.10), $f'(k^\circ) = \beta$, whence $k^\circ = f'^{-1}(\beta)$ where f'^{-1} stands for the inverse function of f'. The balance equation implies $c^\circ = f(k^\circ)$. In view of the optimum condition, if $k > k^\circ$, then $f'(k) - \beta < f'(k^\circ) - \beta = 0$, hence $\dot{c} < 0$. Similarly, $k < k^\circ$ implies $\dot{c} > 0$.

The process can also be analyzed graphically (Kamien and Schwartz, 1981, Figures 17.6 and 17.7).

The set of points (k, c), satisfying condition $\dot{c} = 0$, is (k°, c) where c is an arbitrary positive number, that is, the set is a vertical line. To the left from it consumption increases, to the right from it consumption decreases (Figure 10.1).

The set of points (k, c), satisfying condition $\dot{k} = 0$, is defined by $c = f(k)$ where k is an arbitrary positive number. Because of our assumptions concerning the production function, this curve is degressively increasing (concave). Above it the stock increases, below it the stock decreases. ∎

In Figure 10.2 four paths are considered. Path 1 is the phase curve of the optimal accumulation path corresponding to $k_0 < k^\circ$. If we choose higher consumption (Path 2), then sooner or later accumulation is halted, moreover, the capital is run through. If we choose lower consumption (Path 3), then sooner or later consumption decreases. Then by a sudden, discontinuous increase in consumption, we can jump to Path 4, to the phase curve of the optimal accumulation path corresponding to $k_0 > k^\circ$.

First we shall analyze the local dynamics. We *linearize* the balance equation (10.23b):

$$(10.24) \qquad \dot{k} = f'(k^\circ)(k - k^\circ) - (c - c^\circ)$$

and the optimum-condition (10.23a):

$$(10.25) \qquad \dot{c} = (a^\circ)^{-1} f''(k^\circ)(k - k^\circ)$$

where a° is the absolute risk aversion coefficient of consumption at steady state. Note the similarity between (10.11) and (10.25).

Taking the derivative of (10.24), and plugging (10.25) into the new equation, we obtain a second-order differential equation:

$$(10.26) \qquad \ddot{k} = f'(k^\circ)\dot{k} - \dot{c} = f'(k^\circ)\dot{k} - (a^\circ)^{-1} f''(k^\circ)(k - k^\circ).$$

The solution of (10.26) is determined by the corresponding characteristic polynomial (Example 5.10):

$$\lambda^2 - f'(k^\circ)\lambda + (a^\circ)^{-1} f''(k^\circ) = 0.$$

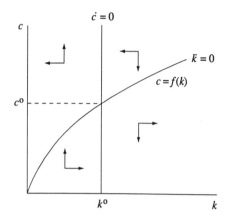

Figure 10.1
Phase plane and niveau lines

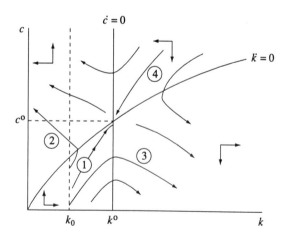

Figure 10.2
Saddle-point and instability

Since $f'(k^o) > 0$, $f''(k^o) < 0$ and $a^o > 0$, the first root (λ_1) of the quadratic equation is positive, the second (λ_2) is negative, thus (k^o, c^o) is a saddle point. The general solution is given by

$$k(t) = k^o + k_1 e^{\lambda_1 t} + k_2 e^{\lambda_2 t}.$$

The solution is convergent if and only if $k_1 = 0$, that is, $k(t) = k^o + k_2 e^{\lambda_1 t}$. $k(0) = k_0$ implies $k_2 = k_0 - k^o$.

We have proved

Theorem 10.4. *The steady state of the optimal capital accumulation is locally saddle-point stable. A path is stable if and only if the initial condition satisfies $k_1 = 0$.*

The nonlinear analysis is more difficult but its result is similar. Note that the proof in Appendix A, Chapter 2 of Blanchard and Fischer (1989) is erroneous. Burmeister (1981) and Stokey and Lucas (1989) gave a thorough discussion for the discrete-time case.

We shall present two problems to be solved, which illustrate the general discussion.

Problem 10.2. Cobb–Douglas production and CRRA utility functions. Let $f(k) = Ak^\alpha$ and $u(c) = \sigma^{-1} c^\sigma$. Determine the formulas for k^o, c^o, r^o and λ_1.

Problem 10.3. Numerical illustration. Let $A = 10$, $\alpha = 0.3$ and $\sigma = -2$ in Problem 10.2. Calculate k^o, c^o, r^o and λ_1.

Here we mention an outstanding result of the neoclassical growth theory.

Theorem 10.5. *(Cass, 1965 and Koopmans, 1965.) For the infinite horizon capital accumulation problem, the optimal path converges to the golden rule steady state.*

Problem 10.4. Let $A = 10$, $\alpha = 0.3$, and $\sigma = -2$. Calculate with numerical integration (using annual steps) the optimal paths for various c_0.

Decentralized solution

It can be shown that the optimum of the centralized problem can be achieved in a decentralized way (Blanchard and Fischer, 1989, Section 2.2). There are two markets for the factors of production: one for labor and another for capital. Wage is $w(t)$, the rent of capital, that is, interest rate is $r(t)$.

For simplicity's sake, we assume that there are a lot of families with identical utility functions. Every family decides on the quantity of labor and capital it lends to the firms, on the quantity it saves and consumes. There are a lot of firms with identical production functions. They produce with the help of rented labor and capital. Assuming constant returns to scales and perfect competition, the number of firms is indifferent.

We assume that the families as well as the firms have perfect foresight. That is, for given paths $\{w(t), r(t)\}$, every family maximizes its remaining life-time utility at date τ, $I[c] = \int_\tau^\infty e^{-\beta t} u[c(t)] \, dt$ under the continuum of budget constraints

$$c(t) + \dot{k}(t) = w(t) + r(t)k(t).$$

Each firm maximizes its profit. According to well-known conditions,

$$f'[k(t)] = r(t) \quad \text{and} \quad f[k(t)] - r(t)f'[k(t)] = w(t).$$

It can be shown that the decentralized optimum of families and firms is identical to the centralized optimum of the planner.

Remarks.   1. What happens if the consumer or the producer has an imperfect forecast on wage or interest rate? The optimal solution would be different (see Problem B.4 and Section C.4).

2. In contrast with Blanchard and Fischer (1989, Section 2.3), we excluded private debts. If we add the government, too, then it is suitable to add private and government debt to the model. Then one can discuss the influence of different taxation systems on the optimal consumption path.

10.3 CONSTANT CAPITAL-OUTPUT RATIO: FINITE HORIZON

In the previous Section we analyzed the case of infinite horizon to get rid of the complexities of terminal conditions. We turn now to the case of finite horizon. Following Tinbergen (1960), we shall discuss the optimal consumption path on a finite horizon, assuming constant output-capital ratio (A):

$$f(k) = Ak, \qquad A > 0.$$

Reinterpreting the problem in Section 10.1, $k(t)$ is now productive capital rather than accumulated savings to be exhausted. There are no external wages and the interest rate is equal to the output-capital ratio, while investment is equal to saving ratio times output:

$$w(t) = 0, \quad r = A \quad \text{and} \quad I = sY.$$

We have the analogy of Theorem 10.1:

Theorem 10.6. *Assuming that the viability conditions* $0 < s(t) < 1$ *are satisfied, the optimal consumption path is as follows:*

$$c(t) = c_0 e^{\gamma t}$$

where the growth rate of consumption is

(10.27)
$$\gamma = \frac{A - \beta}{\zeta}$$

and the initial consumption is

(10.28)
$$c_0 = \frac{(A - \gamma)(e^{AT} k_0 - k_T)}{e^{AT} - e^{\gamma T}}.$$

Remarks. 1. It is evident that for given initial capital k_0, the closing condition k_T also plays an important role in the determination of the optimal path. This might disturb some purists, who can add $G(k_T)$ to objective function (10.22) as was done in (7.1):

(10.22*)
$$I^*[k] = G(k_T) + \int_0^T e^{-\beta t} u[f(k(t)) - \dot{k}(t)] \, dt$$

where $G(\cdot)$ is a smoothly increasing function. We shall retain, however, the original formulation of calculus of variations.

2. In real life, the saving ratio cannot be higher than $1/2$, thus our viability conditions are too mild.

Viability conditions

Until now we skipped the viability conditions $0 < s(t) < 1$. Because of

(10.29)
$$s(t) = 1 - \frac{c(t)}{Ak(t)},$$

the path of $k(t)/c(t)$ is decisive. Look at first the formula for stock:

$$(10.30) \qquad k(t) = e^{At}k_0 - \frac{e^{At} - e^{\gamma t}}{e^{AT} - e^{\gamma T}}(e^{AT}k_0 - k_T).$$

To ensure viability, we need three assumptions.

(i) The relative risk aversion coefficient is higher than unity: $\zeta > 1$ ($\sigma < 0$).

(ii) The discount rate is lower than the output-capital ratio: $\beta < A$.

(iii) The closing stock is higher than the initial stock but it is lower than the maximum achievable by zero consumption. More precisely,

$$(10.31) \qquad 0 < k_0 < \frac{(A - \gamma)e^{\gamma T}}{Ae^{AT} - \gamma e^{\gamma T}}e^{AT}k_0 < k_T < e^{AT}k_0.$$

Remark.   We underline that, apart from the lower bound on k_T in (10.31), every assumption is natural. Assumption (i) asserts that the intertemporal substitutability of consumption is limited, for example, no optimal consumption path contains zero consumption point. (We shall see that accepting Frisch's data from 1931, Tinbergen chose just the opposite assumption and he could not accommodate with his own paradoxical results.)   Assumption (ii) is of empirical nature, $\beta < 0.1 < A$, hence $\beta < A$. The upper bound in Assumption (iii) excludes targets that require zero or negative consumption. The lower bound in Assumption (iii) is quite complex and is derived from the proof below. For better understanding, we note that $\min k_T$ lies between k_0 and the golden rule value $k^* = k_0 e^{\gamma T}$, and the value of k^* depends on γ and T.

We can formulate the completion of Theorem 10.6.

Theorem 10.7. *Let Assumptions (i)–(iii) hold.*

a) The optimal consumption path is a local optimum of the calculus of variations problem.

b) The optimal consumption is increasing in time.

c) The larger the discount rate or the relative risk aversion coefficient (or the lower the elasticity of intertemporal substitution), the higher is the initial consumption and the lower is the growth rate of consumption.

d) If $k_T = k^*$, then the optimal $s(t)$ is constant. If $k_T > k^*$, then $s(t)$ is increasing; if $k_T < k^*$, then $s(t)$ is decreasing.

Proof.   a) If there exists an optimum, then it satisfies the conditions. We should, however, prove the positivity condition. We ought to substitute (10.29) into $0 < s(t) < 1$, but it would be rather cumbersome. Rather, we shall study other, equivalent conditions.

Positivity of c_0. On the R.H.S. of (10.28) the fraction is always positive. $c_0 > 0$ if and only if $k_T < e^{AT}k_0$.

Positivity of $\dot{k}(t)$. Take the derivative of (10.30).

$$\dot{k}(t) = Ae^{At}k_0 - \frac{Ae^{At} - \gamma e^{\gamma t}}{e^{AT} - e^{\gamma T}}(e^{AT}k_0 - k_T).$$

We take the common denominator and summarize the terms on the R.H.S., then write down condition $\dot{k}(t) > 0$. Dropping the positive denominator,

$$A(k_T - e^{\gamma T}k_0)e^{At} + \gamma(e^{AT}k_0 - k_T)e^{\gamma t} > 0.$$

If we are content with a sufficient condition, then inequality $k_T > e^{\gamma T}k_0$ is satisfactory, making the first term positive, while the second term is also positive, as $k_T < e^{AT}k_0$.

If we look for an exact condition, then we had to continue with the calculations. Rearranging the last displayed inequality,

$$(Ae^{At} - \gamma e^{\gamma t})k_T > (Ae^{At+\gamma T} - \gamma e^{AT+\gamma t})k_0 > 0,$$

that is,

$$k_T > \frac{Ae^{At+\gamma T} - \gamma e^{AT+\gamma t}}{Ae^{At} - \gamma e^{\gamma t}}k_0.$$

Let $g(t)$ stand for the fraction, figuring as the multiplier of k_0. Simplifying with $e^{\gamma t}$, yields

$$g(t) = \frac{Ae^{(A-\gamma)t+\gamma T} - \gamma e^{AT}}{Ae^{(A-\gamma)t} - \gamma}.$$

Taking the derivative of function $g(t)$, it can be shown that $\dot{g}(t) > 0$, that is, $g(t) < g(T)$ if $0 < t < T$. Hence the necessary and sufficient condition of viability is $k_T > g(T)k_0$. Using the form of function $g(t)$, the lower bound in (10.31) is obtained.

Positivity of $k(t)$: it follows from $k_0 > 0$ and $\dot{k}_0 > 0$.

b) Since (10.27) and $\beta < A$, $\gamma > 0$ holds.

c) See (10.27).

d) See former proofs.   ∎

Numerical illustrations

We can illustrate the qualitative results on a simple numerical example. Already Tinbergen was disturbed by the difficulty of calibration of the model. We start from two elementary relations; (6.6): $\Gamma = sA$ and (10.27): $\gamma = (A - \beta)/\zeta$. Beginning with an annual growth rate $\Gamma = \gamma = 0.02$ and saving ratio 0.16, the resulting $A = \Gamma/s = 0.02/0.16 = 0.125$ is a too low value. Working with discount rates of several per cents, the resulting intertemporal elasticity of substitution becomes quite large. Striving for rounded numbers, $\beta = 0.065$ yields $\zeta = 3$. With $k_0 = 1/A = 8$ the initial net output is unity: $y_0 = 1$. Finally, for a constant saving ratio $s = 0.16$ and $T = 10$ years, the closing golden rule stock $k^* = e^{\gamma T} k_0 = 9.77$.

Table 10.1 contains the various sets of parameters and characteristics of the runs.

Table 10.1. *Parameters and characteristics*

	Discount rate β	End capital k_T	Growth rate γ	Consumption initial c_0	Consumption end c_T	Saving ratio initial s_0	Saving ratio end s_T
Runs							
1	0.065	9.77	0.02	0.840	1.026	0.160	0.160
2	0.035	9.77	0.03	0.806	1.087	0.194	0.110
3	0.065	10.75	0.02	0.795	0.971	0.205	0.277
4	0.035	10.75	0.03	0.762	1.029	0.238	0.234

Run 1 is the benchmark, the other runs are compared to it: $\beta_1 = 0.065$. In Run 2 the discount rate is almost halved: $\beta_2 = 0.035$, but we retain the previous closing stock: $k_T(2) = k^*$. Then the growth rate of consumption jumps to 3 per cent but the initial consumption drops by 3.4 per cent points, while the closing consumption increases by 6 per cent points.

Run 3 returns to the high discount rate: $\beta_3 = 0.065$ and also to the low growth rate of consumption: 2 per cent. At the same time we raise the closing stock by 10 per cent: $k_T(3) = 10.75$. The saving ratio starts even from a higher value than Run 2, and it is increasing to 27.7 per cent, while the consumption path lags behind both previous cases.

Run 4 unifies the improvement of Runs 2 and 3: the discount rate is diminished and the closing stock is raised. It is true that the initial consumption drops by a further 3.3 per cent points compared to

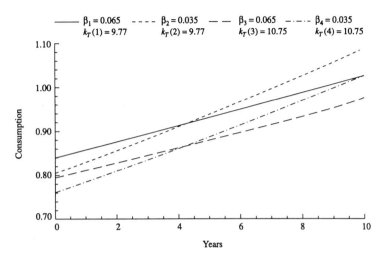

Figure 10.3
Optimal consumption paths

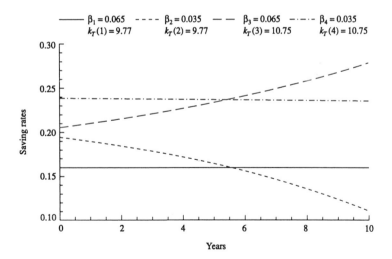

Figure 10.4
Optimal investment paths

Run 3, but the terminal consumption already reaches its counterpart in Run 1. Meanwhile the saving ratio is stabilized at the mean value of Run 3.

As can be seen, even the improved model is far from being perfect. If we accept the results of the modelling, then we can state: in our situation, the discount rate should be diminished and the closing stock should be raised. With due reservations, on the basis of the model the following statements can be made concerning growth policy: 1) If it is possible to diminish the shortsightedness of the decision makers (represented by strong discounting in the model), then diminished initial consumption yields higher growth rate of consumption. 2) If it is possible to reduce the burden on future generations (represented by low closing stock in the model), then the diminishing time trend of saving ratio can be mitigated. We emphasize, however, that there is no mathematical method for the objective comparison of different optimal paths, we have to use our own common sense.

Remarks. 1. As was already mentioned, Tinbergen assumed elastic intertemporal substitution (weak concavity). It is of interest that he introduced the subsistence minimum, under which the consumption cannot drop. As a realistic economist, he went over the problems of order of magnitudes and was dissatisfied with the results. Koopmans (1965) operated with another, probably better assumption: the consumer's marginal utility goes to $-\infty$ while consumption goes to zero (strong concavity). As is seen in Table 10.2, he and Cass (1965) used concave production function and infinite time horizon.

2. Other researchers were less cautious with the choice of their utility functions. For example, a classic of optimal growth theory, Shell (1967a) worked with a linear utility function. Applying the rather new tools of optimal control theory (Pontryagin et al., 1961) he obtained the following result: it is optimal to save (invest) all output in the first part of the period and consume all output plus capital in the second. He was so little interested in the quantitative details that he did not constrain the range of saving ratios from [0,1] to, say, [0.1;0.2]. Is it possible that he wanted to present the difficulty of calculus of variations in the most unrealistic case?

In Hungary, similar results were obtained by Virág (1969), Kovács and Virág (1981), Banai and Lukács (1987). The characteristics of the foregoing models are summarized in Table 10.2.

Table 10.2. *Models and Assumptions*

Assumptions Models	Horizon	Production functions	Utility functions
Tinbergen (1960)	finite	linear	weakly concave
Cass (1965)	infinite	nonlinear	strongly concave
Koopmans (1965)	infinite	nonlinear	strongly concave
Shell (1967)	finite	linear	linear
Virág (1969)	finite	linear	linear
Kovács–Virág (1981)	finite	linear	linear
Banai–Lukács (1987)	finite	ad hoc	linear
Simonovits (1995b)	finite	linear	strongly concave

Further results

We shall need the discrete-time version of Section 10.1 in the analysis of Overlapping Cohorts (Appendix C). The results of Section 10.2 were applied by Blanchard and Fischer (1989) to discuss the problem of deficit-finance and open economy.

It is noteworthy that the optimal solution is *time-consistent* for time-invariant discount rate, that is, at date T_0 replanning the problem for the remaining interval $[T_0, T]$ with the optimal stock $k^o(T_0)$ as initial value, the resulting optimum in $[T_0, T]$ coincides with the original optimum. (Strotz (1955) demonstrated, however, that for time-variant discount rates this is not the case.) Note that our stochastic optimization problems in Section 10.1 also suffer from this issue: for example, in the utility function (10.14) the survival probability function $Q(t)$ is an implicit discount factor and it is not an exponential function. Thus solving the problem again, after τ time elapsed, dynamic inconsistency arises.

PART III

SUPPLEMENTS

This Part completes the book. It consists of three Appendices.

Appendix A summarizes the mathematical theorems of Linear Algebra which are used in the book.

Appendix B outlines *Overlapping Generations* models, where everybody lives exactly two time-periods and in his second period, gives birth to the young. It is a nice application of nonlinear analysis of Chapter 3 and it supplements the results of Sections 8.1 and 8.2 by showing that the equilibrium is not always Pareto-optimal.

Appendix C analyzes *Overlapping Cohorts* models where everybody lives more than 2 time-periods. In contrast to Appendix B, now the existence, uniqueness and stability of the steady state(s) become uncertain and we can only exclude the existence of 2-cycles.

APPENDIX A. LINEAR ALGEBRA

In Appendix A we present some important concepts and theorems from Linear Algebra which have applications in this book but lack for dynamic interpretation. Elementary concepts and theorems like independence are omitted. Useful information can be found in Gantmacher (1959), Varga (1962), Lancaster (1969), Young (1971) and Bermann and Plemons (1979).

Transformations and matrices

Transformation M of $\mathbf{R}^n$ is called *homogeneously linear* if for any pair of vectors x and y and any pair of scalars α and β,

$$(A.1) \qquad M(\alpha x + \beta y) = \alpha M x + \beta M y.$$

For a given system of coordinates, there are n unit vectors with coordinate-forms $e_i = (0, \ldots, 1, \ldots, 0)^{\mathbf{T}}$, where 1 stands at place i, $i = 1, \ldots, n$. In this system of coordinates, transformation M and vector x can be represented by matrix $M = (m_{ij})$ and n-tuple $x = (x_1, \ldots, x_n)^{\mathbf{T}}$, respectively. We need not distinguish transformation and matrix as we do not differentiate between a vector without coordinates and that of coordinates.

Block-structure of matrices

The *block-structure* of matrix M often helps the analysis, making possible the breakdown of larger problems into smaller ones. Let us assume that there is a partitioning of the index set $\{1, 2, \ldots, n\}$ into P disjoint nonempty sets J_Q, $Q = 1, \ldots, P$. By a simultaneous change in rows and columns, we can achieve that the elements of any block be subsequent integers: $J_Q = \{j_{Q-1}, \ldots, j_Q - 1\}$, $Q = 1, \ldots, P$ where $j_0 = 1$ and $j_P = n+1$. Obviously, M has the following *block-structure*:

$$M = (M_{Q,R}) \quad \text{where} \quad M_{Q,R} = (m_{i,j})_{i \in J_Q, j \in J_R}.$$

Assume that M has the following *block-diagonal form*: $M_{QR} = 0$ if $R \neq Q$, that is, $M = \langle M_{QQ} \rangle$. Do partition vector x in a consistent way: $x = (x_Q)$ where $x_Q = (x_i)_{i \in J_Q}$.

The following theorem refers to the role of the block-diagonal transformations:

Theorem A.1.   *The block-diagonal linear transformation M breaks up into P independent linear transformations M_{QQ}, each defined on the corresponding invariant subspace $\mathbf{R}^{n_Q}$, $Q = 1, \ldots, P$.*

Characteristic roots and characteristic vectors

A *characteristic root* (or *eigenvalue*) and a corresponding nonzero *(right hand) characteristic vector* (or *eigenvector*) of transformation M are a scalar λ and a column vector $s \neq 0$, respectively, such that under the transformation the direction of s remains invariant and its magnitude is multiplied by λ:

(A.2) $Ms = \lambda s, \quad s \neq 0.$

It is obvious that if s is a characteristic vector, then any nonzero scalar ($\alpha \neq 0$) multiple of it is also a characteristic vector with the same characteristic root: $M(\alpha s) = \lambda(\alpha s)$. It is known from the theory of homogeneous linear equations that concerning the roots λ, (A.2) is equivalent to $\det(\lambda I - M) = 0$ where 'det' stands for the determinant. Letting λ be a variable, we obtain the *characteristic polynomial* of transformation M:

(A.3) $P(\lambda) = \det(\lambda I - M).$

The next theorem states the main properties of the characteristic polynomial.

Theorem A.2. *The characteristic polynomial $P(\lambda)$ of matrix M is an n-degree polynomial with n real or complex roots with multiplicity. These roots are the characteristic roots in (A.2), denoted by λ_j, $j = 1, \ldots, n$.*

If all the characteristic roots are different, then the corresponding characteristic vectors are linearly independent (Ralston, 1965, Theorem 10.9). It may happen that a characteristic root λ has an *algebraic multiplicity* $r > 1$ in the characteristic equation: $P(\lambda) = P'(\lambda) = \cdots = P^{(r-1)}(\lambda) = 0 \neq P^{(r)}(\lambda)$. Let r^* be the number of linearly independent characteristic vectors belonging to λ, r^* being the *geometrical multiplicity* of λ. Obviously, $1 \leq r^* \leq r$.

It is known that *typically* $r^* = r = 1$, that is, the matrix has a *simple structure* and there exists a basis formed by n linearly independent characteristic vectors, called *characteristic basis*.

We shall now introduce a triple of definitions. The *spectral radius* of a matrix is the maximum of the moduli of the n characteristic roots, its notation is $\rho(M)$: $\rho(M) = \max\{|\lambda_1|, \ldots, |\lambda_n|\}$. A *dominant (characteristic) root* of a matrix is a characteristic root with the maximal modulus. A *dominant (characteristic) vector* of a matrix is a characteristic vector belonging to a dominant characteristic root. A matrix may have multiple dominant characteristic roots and a dominant characteristic root may have multiple normalized characteristic vectors.

Example A.1. Multiple dominant roots. Let α and β be negative scalars. If

$$M = \begin{pmatrix} 0 & \alpha \\ \beta & 0 \end{pmatrix},$$

then both $\rho(M)$ and $-\rho(M)$ are dominant roots and their modulus is equal to $\sqrt{\alpha\beta}$.

We shall also need the concept of *left hand characteristic vector*:

(A.4) $$p^{\mathbf{T}}M = \lambda p^{\mathbf{T}}, \quad p \neq 0.$$

Going back to the right hand characteristic vectors, (A.4) is equivalent to $M^{\mathbf{T}}p = \lambda p$ where $M^{\mathbf{T}}$ is the *transpose* of matrix M: $m_{ij}^{\mathbf{T}} = m_{ji}$, $1 \leq i, j \leq n$. Since the characteristic polynomial of $M^{\mathbf{T}}$ is identical to that of M, there is no need to make distinction between left and right hand characteristic roots.

The following theorem describes an interesting connection between left and right hand characteristic vectors associated with different roots.

Theorem A.3. *Left and right hand characteristic vectors with different characteristic roots are orthogonal to each other.*

Proof. Let $\varepsilon p^{\mathbf{T}} = p^{\mathbf{T}}M$ and $\lambda s = Ms$, $p \neq 0 \neq s$, $\lambda \neq \varepsilon$. Multiplying the first equation from the right by s and the second from the left by $p^{\mathbf{T}}$, yields $\varepsilon p^{\mathbf{T}}s = p^{\mathbf{T}}Ms = \lambda p^{\mathbf{T}}s$, hence $p^{\mathbf{T}}s = 0$. ∎

Cayley–Hamilton Theorem*

It is obvious that the space of the linear transformations of $\mathbf{R}^n$ is also linear with dimension n^2. This implies that the matrices of sequence $I, A, A^2, \ldots, A^{n^2-1}$ are linearly dependent. However, there is a very important sharpening of this elementary result:

Theorem A.4. *(Cayley–Hamilton theorem, see Lancaster, 1969, Section 4.3.) Every square matrix M satisfies its own characteristic equation:* $P(M) = 0$.

Instead of proof. The general proof is a little bit involved, but for matrices with simple structure, the proof is almost trivial. Let λ_j and s_j be the jth characteristic root and vector, respectively. Then the characteristic polynomial is $P(\lambda) = (\lambda - \lambda_1)\cdots(\lambda - \lambda_n)$, thus $P(M) = (M - \lambda_1 I)\cdots(M - \lambda_n I)$; since factors $M - \lambda_j I$ commute with each other, $P(M)s_j = 0$ for every j, hence $P(M) = 0$. ∎

On the basis of Theorem A.4, for any t, $\{M^t\}$ can easily be calculated from the first $n - 1$ powers of M. Let $P(\lambda) = 0$, or $\lambda^n = \sum_{j=0}^{n-1} \pi_j \lambda^j$. Then $M^n = \sum_{j=0}^{n-1} \pi_j M^j$. Multiplying this formula by M, $M^{n+1} = \sum_{j=0}^{n-1} \pi_j M^{j+1}$ and using the formula for M^n again, yields $M^{n+1} = \sum_{j=0}^{n-1}(\pi_{j-1} + \pi_{n-1}\pi_j)M^j$, $(\pi_{-1} = 0)$ and so on.

Jordan normal form*

For problems with special structure, a characteristic basis may not exist.

Example A.2. Geometrical multiplicity is less than algebraic multiplicity. Consider an r-dimensional matrix

$$
\text{(A.5)} \qquad N_r = \begin{pmatrix} 0 & 1 & 0 & \cdots & 0 \\ 0 & 0 & 1 & \cdots & 0 \\ \vdots & \vdots & \vdots & \ddots & \vdots \\ 0 & 0 & \cdots & \cdots & 1 \\ 0 & 0 & 0 & \cdots & 0 \end{pmatrix}.
$$

Since $N_r^r = 0$, all characteristic roots of matrix N_r are zero, and apart from the scalar multiples, the only characteristic vector is $s = (1, 0, \ldots, 0)^{\mathbf{T}}$.

The following theorem states that, apart from the diagonal matrices, in essence, (A.5) is the simplest form a matrix can take.

Theorem A.5. *(Jordan normal form, Lancaster, 1969, Section 4.13.)* With a suitable similarity transformation S, any transformation M can be transformed into a a block-diagonal form $M^* = S^{-1}MS$ where each block has the normal form $M_Q^* = \lambda_Q I_{r_Q} + N_{r_Q}$ [(A.5)], $Q = 1, \ldots, P$.

In a geometrical language, in each block there are r_Q *principal vectors* denoted by $s_{j-1} = (\lambda I - M)s_j$, $j = r_Q, r_Q - 1, \ldots, 1$ where s_1 is the single genuine characteristic vector and $s_0 = 0$. These vectors form a basis for the invariant subspace of M_Q (Halmos, 1958).

Matrix exponent

In Section 5.3 the idea of defining the exponent of a matrix arose. Fortunately, the Taylor series of e^z is defined for any complex z, thus, at least formally, it can also be defined for any square matrix M.

Theorem A.6. *(Lancaster, 1969, Section 5.6).* For any $n \times n$ matrix M, the matrix exponent can be defined by the Taylor series of e^z:

$$
\text{(A.6)} \qquad e^M = \sum_{k=0}^{\infty} \frac{M^k}{k!}.
$$

In the same way as with (real or complex) scalars, the R.H.S. of (A.6) absolutely converges in entries as well as in norms.

For a block-diagonal matrix, the matrix exponent is equal to the matrix formed from the block exponents: $e^{\langle M_Q \rangle} = \langle e^{M_Q} \rangle$. Specifically, for a diagonal matrix $M = \langle m \rangle = \langle m_j \rangle_j$, the matrix exponent is equal to the diagonal matrix of the exponents: $e^{\langle m \rangle} = \langle e^{m_j} \rangle_j$. For a matrix with a characteristic basis, e^M is similar to $\langle e^{\lambda_j} \rangle_j$.

Remarks. 1.* With the help of the Jordan normal form it is possible to describe $e^{(\lambda I + N)t}$ in a relatively simple form.

2. In connection with Theorem A.4, we have already seen that the calculation of M^k can be reduced to those of $I, M, M^2, \ldots, M^{n-1}$. Therefore we can determine the exponential function of any matrix without using Taylor series of matrices. This procedure carries over to any analytic function, under the assumption that its convergence radius is larger than the spectral radius of the matrix.

Nonnegative and irreducible matrices

Obviously, *matrices with positive entries* (shortly, *positive matrices*) are very important in economics. Formally, $m_{ij} > 0$ for $i, j = 1, \ldots, n$; or $M > 0$.

As a generalization, we also consider *matrices with nonnegative entries* (shortly, *nonnegative matrices*). Formally, $m_{ij} \geq 0$ for $i, j = 1, \ldots, n$; or $M \geq 0$.

Sometimes block-diagonal matrices are considered, because they lead to smaller-size matrices. In other cases, our aim is just the opposite, namely, we want to avoid any further partitioning. Then the following concept is suitable.

Matrix M is called an *irreducible* (or *indecomposable*) matrix if the index set $\{1, 2, \ldots, n\}$ cannot be partitioned into nonempty sets J and $\bar{J}$ in such a way that at least one of the off-diagonal block matrices $M_{J\bar{J}}$ and $M_{\bar{J}J}$ is zero.

To interpret this notion less formally, imagine that we have a *directed graph* of n vertices and a directed edge connects vertex j to vertex i if and only if entry (i, j) is positive. Then irreducibility means that it is impossible to divide the vertices in two groups in such a way that the vertices of the first group are not connected to those of the second group or vice versa. Of course, a matrix is called *reducible* if it is not irreducible. (Note that a block-diagonal matrix is reducible, since for any partitioning, neither group is connected to the other.) The following theorems were discovered between 1907 and 1912 by

Perron (on positive matrices) and Frobenius (on nonnegative matrices). Since the 1950's these theorems have been playing a fundamental role in a lot of economic applications of matrix theory.

Theorem A.7. *(Frobenius 1, see Lancaster, 1969, Theorem 9.2.1.) Let matrix M be nonnegative and irreducible. Then*

a) M has a positive characteristic root λ_1 equal to the spectral radius of M, a dominant root.

b) Apart from a scalar, there exists a unique positive right hand characteristic vector s_1 and it is associated with λ_1.

c) The characteristic root λ_1 has an algebraic multiplicity 1.

d) (Gantmacher, 1959, Lemma 13.2.) The positive dominant root is an increasing function of any entry of the matrix.

e) If the spectral radius of M is less than 1, then $(I - M)^{-1}$ exists and is positive.

We only refer to the basic idea of the proof: Consider function

$$\rho(x) = \min \left[\frac{(Mx)_i}{x_i} \quad \text{where} \quad 1 \leq i \leq n, \quad x_i \neq 0 \right].$$

On the unit sphere $|x| = 1$, function $\rho(x)$ attains its maximum $\rho(M)$ at $x = s_1 > 0$. ∎

Problem A.1. Prove Theorem A.7 for $n = 2$ directly.

Examples A.1–A.2 show that Theorem A.7 is not true for non-negative reducible matrices.

To sharpen points a)–c) of Frobenius' Theorem 1, we shall introduce the following definition. Matrix M is called *P-cyclic* (or *imprimitive* with order P), if $M_{QR} = 0$ for $R \neq Q+1$, that is, $M = (M_{Q,Q+1})$: (1.51). Using the graph interpretation given above: the vertices can be partitioned into P distinct groups in such a way that group 1 is only connected to group 2, group 2 is only connected to group 3,..., group $P - 1$ is only connected to group P, and group P is only connected to group 1. Note that then M^P is a block-diagonal matrix. If M is not a cyclic matrix for any P, then we shall use the adjective *acyclic* (or *primitive*), and write $P = 1$.

Example A.3. 2-cyclic matrix

$$(A.7) \qquad\qquad N = \begin{pmatrix} 0 & N_1 \\ N_2 & 0 \end{pmatrix}$$

where N_1 and N_2 are $r \times (n - r)$ and $((n - r) \times r)$-size matrices, respectively, $0 < r < n$.

Theorem A.8. *(Frobenius 2, Lancaster, 1969, Theorem 9.2.2.) Let the square matrix M be nonnegative and irreducible.*

a) If M is P-cyclic ($P > 1$), then (and only then) M has $P > 1$ dominant characteristic roots $\rho \varepsilon^{Q-1}$, $Q = 1, \ldots, P$, where ε is the P-order complex unity root.

b) If M is acyclic ($P = 1$), then there exists a unique dominant root which is positive. An appropriate power of the matrix is positive, for example, $M^n > 0$.

Remark. If $M > 0$, then it is acyclic (primitive), that is, b) reduces to Perron's theorem. The next example shows that there are acyclic matrices which are not positive.

Example A.4. Acyclic matrix with zero entry.

$$(A.8) \qquad M = \begin{pmatrix} 1 & \alpha \\ \beta & 0 \end{pmatrix}, \qquad \alpha, \beta > 0$$

is acyclic, since

$$M^2 = \begin{pmatrix} 1 + \alpha\beta & \alpha \\ \beta & \alpha\beta \end{pmatrix} > 0.$$

Problem A.2. Prove Theorem A.8 for $n = 2$.

We have seen in Theorem A.7e that nonnegative (discrete-time) stable matrices have a nice property. We show another one, called *productivity*: there exists a column vector $x > 0$ such that

$$(A.9) \qquad\qquad Mx < x.$$

Theorem A.9. *If a square matrix M is nonnegative and irreducible, then (discrete-time) stability is equivalent to productivity.*

Proof. Let $\rho > 0$ and $s > 0$ be the dominant characteristic root and vector of M, $Ms = \rho s$. a) If M is stable, then $\rho < 1$, that is, $Ms = \rho s < s$. b) If M is unstable, then $\rho \geq 1$, that is, $Ms = \rho s \geq s$.∎

We have the following

Corollary. *a) Productivity is equivalent to the existence of a row vector* $p^T > 0$, *such that*

(A.10) $$p^T M < p^T.$$

b) By a suitable choice of the units of measurement, productivity is equivalent to

(A.11) $$1^T M^* < 1^T \quad \text{where} \quad 1^T = (1, \ldots, 1).$$

Proof. a) Taking the transpose of M, (A.9) yields (A.10).

b) Using the notations of Theorem A.9, take the *similar* matrix $M^* = \langle p \rangle M \langle p \rangle^{-1}$. Evidently, 1^T is a left hand dominant characteristic vector of M^* with ρ as the dominant root. ∎

We shall now consider a special class of matrices called *M-matrices*, which plays an important role in numerical analysis as well as in economic applications (Young, 1971, Sections 2.6 and 2.7) and (Metzler, 1945). Because of applications, we shall use the notation B rather than M. The diagonal entries of matrix B are positive (unitary) and the off-diagonal entries are nonpositive:

(A.12) $$b_{ii} = 1 \quad \text{and} \quad b_{ij} \leq 0, \quad i \neq j.$$

We shall introduce the *matrix of cross-effects*:

(A.13) $$N = I - B \geq 0,$$

and assume that N is irreducible, moreover, *the own-effects dominate the cross-effects*:

(A.14) $$-\sum_{i \neq j} b_{i,j} < 1, \quad j = 1, \ldots, n.$$

Then (A.14) is the same as (A.11).

Discussing the stability of continuous-time systems, we need the following

Theorem A.10. *All the characteristic roots of an M-matrix have positive real parts.*

Problem A.3. Consider a 2×2 matrix, with possibly negative entries. Let both column sums be the same positive number. When is it true that the dominant characteristic vector of the matrix is positive?

Norms of vectors and of matrices

For thorough analysis, the mathematical concepts of vector-norms and matrix-norms are practical (Lancaster, 1969, Chapter 6).

A positive function defined on the linear space $\mathbf{R}^n$ is called a *vector-norm* and is denoted by $||x||$ if the following conditions hold:
(i) $||x|| = 0$ if and only if $x = 0$;
(ii) $||\lambda x|| = |\lambda|\, ||x||$, λ is a scalar (homogeneity);
(iii) $||x + y|| \leq ||x|| + ||y||$ (triangle inequality).

Example A.5. Important norms. (a) The best-known vector-norm is the Euclidean distance: $||x||_2 = \sqrt{\sum_{i=1}^n |x_i|^2} = \sqrt{x^{\mathbf{T}} x}$.

(b) Two other vector-norms (l_1 and l_∞) are also very useful:

$$(A.15) \qquad ||x||_1 = \sum_{i=1}^n |x_i| \quad \text{and} \quad ||x||_\infty = \max_{1 \leq i \leq n} |x_i|.$$

Problem A.4. Draw the unit balls for l_1, l_2 and l_∞, $n = 2$.

As is known from linear algebra, linear transformations of linear spaces also form a linear space. It is suitable to define the *matrix-norm* in accordance with the underlying vector-norm: The *induced norm of a matrix* is the supremum of the vector-norms of the images of the vectors of the unit ball. In formula:

$$(A.16) \qquad ||M|| = \sup\{||Mx|| : ||x|| = 1\}.$$

Because of linearity, for an arbitrary vector x

$$(A.17) \qquad ||Mx|| \leq ||M||\, ||x||.$$

Theorem A.11. *(Lancaster, 1969, Section 6.3.) If M and N are linear transformations on $\mathbf{R}^n$, then the function defined by (A.16) is indeed a matrix-norm on $\mathbf{R}^{n^2}$, that is, it is a vector-norm and $||MN|| \leq ||M||\,||N||$.*

Remark. It is not easy to calculate the norm of a matrix in general. It is obvious that the spectral radius provides a lower bound on the induced norm: $\rho(M) \leq ||M||$. (In fact, if $\lambda x = Mx$, then (A.17) and (ii) imply $|\lambda|\,||x|| = ||\lambda x|| \leq ||M||\,||x||$.)

We have, however, an easy task with the vector-norms in (A.15). Then the induced matrix-norms are respectively

$$(A.18) \qquad ||M||_1 = \max_{1 \leq j \leq n} \sum_{i=1}^n |m_{ij}|, \quad ||M||_\infty = \max_{1 \leq i \leq n} \sum_{j=1}^n |m_{ij}|.$$

Problem A.5. Prove (A.18).

Remark.* In modern mathematics the concept of *complete normed linear space* (the so-called *Banach space*) plays an outstanding role. This space is a (generally infinite dimensional) linear space where a) the distance between two vectors is given by the norm of their difference and b) every Cauchy series has a limit: $\lim_{j,k}(x_j - x_k) = 0$ implies the existence of an x_∞ such that $\lim_k x_k = x_\infty$. One of the most important Banach spaces is the space of continuous functions defined on the interval $[a,b]$, denoted by $C[a,b]$ where the norm is the maximum of the function: $||f|| = \max\{|f(x)|, \quad a \leq x \leq b\}$. In Chapters 5, 7 and 9 we refer to these abstract spaces.

APPENDIX B. OVERLAPPING GENERATIONS

This Appendix investigates the *interaction of two generations* at any given time: the young and the old trade with each other. We shall confine our attention to a closed exchange economy where output is equal to consumption and there is neither capital accumulation nor output storage. This Appendix is a simplified and revised version of Gale (1973, Part I).

The model

Following the *life-cycle model* of Modigliani and Brumberg (1954), Samuelson (1958) modelled the dynamic interaction between two or three *overlapping generations*, for short, OLG. By now, OLG has become a central part of macroeconomics. This expansion is reflected in most advanced macroeconomic textbooks (for example, Blanchard and Fischer, 1989; Azariadis, 1993), in which the family of OLG models plays a decisive role.

In each time-period, every (unisex) old, born in the previous time-period, gives birth to ν young. (Of course, in OLG models *the population growth factor ν* can be any positive real number. In reality, the number of children of each couple is an integer (what about divorces?) and only the combination of families with different sizes can produce a noninteger 2ν.)

This Appendix studies an *exchange economy*. We neglect the production process, (apart from Remark 2 to Theorem B.1) we do not model the increase in endowments (increase in productivity) and take the current incomes (endowments) as given exogenously in each time-period. The (time-invariant) current incomes of the young and of the old are w_0 and w_1, respectively. By normalization, $w_0 + w_1 = 1$.

Similarly, at time-period t the consumption of the young and that of the old are $c_{0,t}$ and $c_{1,t}$, respectively. The consumption path of a person born at date t is $c_{0,t}, c_{1,t+1}$. The series $\{c_{0,t}, c_{1,t+1}\}$ is called *consumption program*. It will be useful to introduce the *per-period savings*: $s_{i,t} = w_i - c_{i,t}$.

Another feature of the exchange economy is that the output is not storable, therefore total income and total consumption are equal, thus total saving is equal to zero. Since ν young correspond to an old, in view of zero-saving assumption, a saving program $\{s_{0,t}, s_{1,t+1}\}$ is called feasible if

(B.1) $$\nu s_{0,t} + s_{1,t} = 0, \quad t = 0, 1, \ldots.$$

Using the standard method of the neoclassical economics, we shall derive consumption paths from the maximization of a well-behaved utility function. For the sake of simplicity, let the utility function be additive in time:

(B.2) $$U(c_{0,t}, c_{1,t+1}) = u(c_{0,t}) + v(c_{1,t+1}).$$

It is quite customary, especially for the case of many cohorts, that the *felicity functions* only differ in a scalar, called *discount factor*. In our 2-period case:

$$v(c) = \beta u(c), \quad 0 < \beta \leq 1.$$

In Chapters 8 and 10 we discussed the CRRA utility function in detail. Here we have $u(c) = \sigma^{-1} c^{\sigma}$, $\sigma \neq 0$ or $u(c) = \log c$ ($\sigma = 0$) *Cobb–Douglas utility function*. It is very simple and often realistic to work with *Leontief utility function*: $U(c_{0,t}, c_{1,t+1}) = \min\{c_{0,t}, c_{1,t+1}\}$. In the Leontief case, the conditional optimum is given by $c_{0,t} = c_{1,t+1}$. Of course, this utility function is not additive but it can be approximated by the transforms of CRRA utility functions at $\sigma \to -\infty$.

For the time being, the *interest factor*(=1+interest rate) of time-period t is taken as given and is denoted by $\{r_t\}$. It is assumed that the bequest at the corresponding interest factor is zero at any time-period:

(B.3) $$r_{t+1} s_{0,t} + s_{1,t+1} = 0.$$

A program is called *competitive* if it is optimal with respect to the given interest factor series. Feasible competitive programs are called *equilibrium programs*.

In the continuous-time models of Chapter 10, r stood for interest rate, on an annual basis its value was close to zero. Like in Section 8, here r denotes the interest factor, on an annual basis its value is close to 1.

How does the model work? Express $s_{1,t+1}$ from (B.3) and, after adding the endowments, substitute the resulting consumption vector into (B.2). Except for the Leontief case, the local maximum can be determined from the following first-order condition:

$$\text{(B.4)} \qquad u'(c_{0,t}) = r_{t+1}v'(c_{1,t+1}).$$

Then we express $s_{0,t}$ and $s_{1,t+1}$ as functions of r_{t+1}: $s_{0,t} = s(r_{t+1})$ and by (B.3), $s_{1,t+1} = -r_{t+1}s(r_{t+1})$, these are the *conditional saving functions*. Using Gale's short-cut, it is assumed that the policy functions are time-invariant, namely, $s_{1,t} = -r_t s(r_t)$. Inserting them into the feasibility condition (B.1), yields the implicit equation

$$\text{(B.5)} \qquad S(r_t, r_{t+1}) = \nu s(r_{t+1}) - r_t s(r_t) = 0.$$

It is not certain that the implicit difference equation has a solution and if it has, then whether it is unique.

We shall occasionally interrupt the flow of general concepts and theorems and discuss several concrete examples.

Example B.1. In the Cobb–Douglas case the young's conditional consumption and saving functions are

$$c(r) = \frac{w_0 + w_1 r^{-1}}{1 + \beta} \quad \text{and} \quad s(r) = \frac{w_0 \beta - w_1 r^{-1}}{1 + \beta}.$$

Problem B.1. Demonstrate that in the Leontief case the conditional consumption functions are

$$c_0(r) = \frac{r w_0 + w_1}{1 + r} = c_1(r).$$

Problem B.2. Using present values, prove for CRRA utility function, that the young's conditional consumption and saving functions are as follows:

$$c_0(r) = \frac{w_0 + w_1 r^{-1}}{1 + \Phi r^{-\mu}} \quad \text{and} \quad s(r) = \frac{w_0 \Phi r^{-\mu} - w_1 r^{-1}}{1 + \Phi r^{-\mu}}$$

where $\mu = \sigma/(\sigma - 1)$, $1 - \mu$ is the *intertemporal elasticity of substitution* and $\Phi = \beta^{1-\mu}$ is the corrected discount factor.

Example B.2. For Cobb–Douglas utility function ($\mu = 0$), $S(r_t, r_{t+1}) = 0$ leads to the difference equation $r_{t+1} = w_1/(w_1 + \nu\beta w_0 - \beta w_0 r_t)$.

Steady states

As is the case in all dynamic models, in this model a special role is played by *steady states* where the subsequent generations' savings paths are identical to each other. Subindex $_F$ refers to a *feasible* steady state:

$$s_{0,t} = s_{0,F} \quad \text{and} \quad s_{1,t+1} = s_{1,F}.$$

We shall see rather soon that the use of this new notation for the steady state makes room for further distinctions.

We shall encounter a special steady state where there is no trade, it is also called *autarky*. In Appendix C we shall generalize this concept to the *balanced steady state* and subindex $_B$ refers to it:

$$s_{0,B} = 0 \quad \text{and} \quad s_{1,B} = 0.$$

It will turn out that the optimal path with a given initial condition is not necessarily stable. In turn, we shall call the optimal steady state *golden rule* and refer to it by subindex $_G$.

With substitution,

$$(B.1°) \qquad \nu s_{0,F} + s_{1,F} = 0,$$
$$(B.3°) \qquad r_F s_{0,F} + s_{1,F} = 0$$

whence $(\nu - r_F)s_{0,F} = 0$. Using the terms introduced above, we arrived at

Theorem B.1. *(Gale, 1973, Theorem 1.) In the OLG exchange economy there exist two steady states: either (i) the golden rule or (ii) the autarky:*

$$\text{either} \quad r_G = \nu \quad \text{or} \quad s_{0,B} = 0 \quad \text{where} \quad r_B = \frac{u'(w_0)}{v'(w_1)}.$$

Remarks. 1. Note that for $r_G = \nu$, the budget constraint (B.3°) coincides with the feasibility condition (B.1°). Therefore only one rather than two constraints are to be taken into account, making this program the optimal steady state, that is, the golden rule.

2. It is widespread to neglect the productivity growth, although this factor frequently dominates the population growth. For convenience, we also follow this custom, but for a moment, we consider the productivity change. Suppose that both earnings are multiplied by

the productivity factor η in every period. Due to the homotheticity of our utility function, the consumption pair also grows by the same growth factor. Then (B.3°) generalizes into $r_F s_{0,F} + \eta s_{1,F} = 0$, whence $(\eta\nu - r_F)s_{0,F} = 0$. That is, the golden rule interest factor is equal to the product of the productivity and the population growth factors: $r_G = \eta\nu$.

Returning to the simple world of unchanged productivity, we shall use the following distinction. The golden rule program of an OLG exchange economy is called *debtor* or *creditor* or *symmetric* if at the golden rule the young dissaves or saves or is just in equilibrium:

$$\text{either} \quad s_{0,G} < 0 \quad \text{or} \quad s_{0,G} > 0 \quad \text{or} \quad s_{0,G} = 0.$$

Gale (1973) spoke of *classical, Samuelson* and *coincidental* golden rule. We prefer the figurative terms by Augusztinovics (1992). (Alternatively, some other authors speak of monetary steady state if the golden rule is debtor and of nonmonetary steady state in case of autarky or more generally, in case of balanced steady state.)

Example B.3. For a discounted Cobb–Douglas utility function, the golden rule is equal to

$$c_{0,G} = \frac{w_0 + w_1\nu^{-1}}{1 + \beta}, \quad c_{1,G} = \frac{\beta + (1 - \nu^{-1})}{1 + \beta}$$

and the autark interest factor is given by

$$r_B = \frac{w_1}{\beta w_0}.$$

The type of the golden rule depends on the relation between the autark interest factor and the population growth factor.

Theorem B.2. *(Gale, 1973, Theorem 2.) The golden rule of the OLG exchange economy is creditor (debtor) if and only if the autark interest factor is less (greater) than the population growth factor:*

(B.6) $r_B < \nu \quad (\text{or} \quad r_B > \nu).$

Proof. Since $r_B \neq \nu$, c_G does not satisfy the individual budget constraint (B.3°), it costs more than that: $r_B s_{0,G} + s_{1,G} < 0$. By (B.1°), $\nu s_{0,G} + s_{1,G} = 0$. By deduction, $(r_B - \nu)s_{0,G} < 0$, and together with the definitions, this implies (B.6)   ∎

We shall present the following Example.

Example B.4. Cobb–Douglas illustration. With the comparison of Problem B.1 and Example B.1, it is easily checked that Theorem B.2 holds. Indeed, $c_0(\nu) = (w_0 + w_1\nu^{-1})/(1+\beta) > w_0$ and $r_B = w_1/(\beta w_0) > \nu$ are equivalent.

Finally, one can formulate

Theorem B.3. *(Gale, 1973, Theorem 3.)* In the OLG exchange economy the autarky is Pareto-optimal if and only if the golden rule is debtor.

Proof.   a) If the golden rule is creditor, then $s_{0,G} > 0$. Therefore program $\{w\}$ can be improved, since we can turn to program $\{c_G\}$.

b) If the golden rule is debtor, then consider a program $\{c_{0,t}, c_{1,t+1}\}$ which is as good as $\{w_0, w_1\}$. Then using the principle applied in the proof of Theorem B.2, $r_B s_{0,t} + s_{1,t+1} > 0$. Inserting condition (B.1) for $t+1$, after rearrangement we obtain $(r_B/\nu)s_0 < s_{0,t+1}$. Repeating the process for $t = 0, 1, 2, \ldots, T-1$, yields $(r_B/\nu)^T s_{0,0} < s_{0,T} < w_0$. By (B.6), $r_B > \nu$, and $s_{0,0} > 0$, hence for $T \to \infty$ we obtain a contradiction. ∎

We remind the reader that in the traditional general equilibrium models (for example, Arrow and Debreu, 1954) the equilibrium is generally Pareto-optimal. In contrast, here this is not usually the case. The most plausible explanation for this anomaly may be that there exist an infinite number of consumers and goods. Going deeper, however, it turns out that the real trouble is different, namely, certain *markets are missing*.

Local analysis

We are ready to analyze the nonstationary programs.

Theorem B.4. *(Gale, 1973, Theorem 4.)* In a debtor OLG exchange economy the autarky is locally unstable, while in a creditor OLG exchange economy the autarky is locally stable.

Proof.   Linearizing (B.3) around the autarky and taking into account that steady states' savings are zeros, we obtain

$$(B.7) \qquad\qquad s_{0,t+1} = \nu^{-1} r_B s_{0,t}.$$

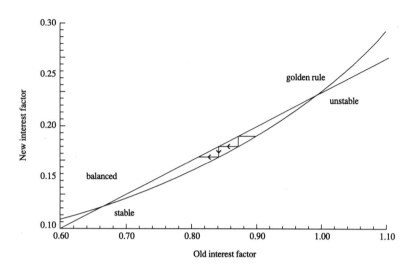

Figure B.1
Interest dynamics in the phase plain: Cobb–Douglas utility function

The scalar version of Theorem 3.3 implies the condition of local stability: $\nu^{-1} r_{\mathrm{B}} < 1$, Together with Theorem B.2, this completes the proof. ∎

Remark. In fact, without giving any proof, Gale 'completed' Theorem B.4 as follows: *Debtor golden rule is locally stable, while creditor golden rule is locally unstable.*

The following problem contains a simple case, with a simple proof for the claim, illustrated in Figure B.1.

Problem B.3. Prove that if function $r_{t+1} = f(r_t)$ is increasing and strictly convex (as is the case for a Cobb–Douglas utility function in Example B.2), then Gale's claim on the stability/instability of the golden rule is correct.

To find the counterpart of Theorem B.4, we introduce the following notations: The young's conditional saving function is $s(r)$, its inverse elasticity at $r = \nu$ is $\varepsilon = -s(\nu)/s'(\nu)$.

Theorem B.5. *The golden rule steady state is locally stable if*

$$(B.8) \qquad\qquad 0 < \varepsilon < 2\nu.$$

Proof. $S(\nu, \nu) = 0$. Using notation ε and the implicit function theorem,

$$\frac{dr_{t+1}}{dr_t} = \frac{-\partial S/\partial r_t}{\partial S/\partial r_{t+1}} = \frac{s(\nu) + \nu s'(\nu)}{\nu s'(\nu)} = 1 - \frac{\varepsilon}{\nu}.$$

Using the stability consition for scalar systems, (B.8) is obtained. ∎

For a moment, we leave the world of rational expectations and consider naive expectations, replacing r_{t+1} with r_t in (B.4) (for details, see Sections 4.5 and C.4)

Problem B.4. Following Gale (1974), assume that the consumer has naive rather than rational expectations. Prove that a) the young's conditional optimum remains invariant, b) the modified difference equation is $S(r_t, r_{t+1}) = \nu s(r_{t+1}) - r_{t+1} s(r_t) = 0$. c) The local stability condition at the balanced state is as before but at the golden rule state either $-\infty < \varepsilon < -2\nu$ or $0 < \varepsilon < \infty$.

Comparing Problem B.4c with (B.8) suggests that Gale's claim may not be true and the stability of rational expectations implies that of naive expectations but not the other way around.

Are these results consistent with optimization? Within the limits of time-additive utility function (B.2), $\varepsilon > 0$, thus the negative values can be ruled out in the modified stability condition. However, if the utility function is nonadditive or there are different classes of agents, then probably *everything goes* (Blanchard and Fischer, 1989, p. 248).

Problem B.5. Using Problem B.2, prove for CRRA utility functions, a) Gale's claim holds and b) the two expectations are equistable.

However, below (see also Sections C.4–C.5) we shall refute the claim with a simple modification of Gale's famous quadratic utility function. At this job, another route to the basic difference equation will be useful. Express r_{t+1} from (B.3) and substitute it into (B.4). The result is

(B.9) $$s_{0,t} u'(c_{0,t}) + s_{1,t+1} v'(c_{1,t+1}) = 0.$$

The implicit difference equation (B.9) describes (not always unambiguously) the series of optimal savings of the young. The consumption path of the old can be determined from (B.1).

Global analysis

We have already mentioned that, in contrast with the one-sector optimal growth theory (Section 8.2), in the OLG exchange economy the equilibrium path may not converge to the steady state. Moreover, it may occur that cyclic and chaotic paths appear (see Chapter 3). Instead of reproducing the usual, quite involved, general argumentation (Benhabib and Day, 1982; Grandmont, 1985), for the time being, we shall be content with the illustration on the simplest examples.

Example B.5. Unstable equilibrium path (generalization of Gale, 1973, Example 3). Let $u(c_0) = 2ac_0 - bc_0^2$ where a and b are positive and let $v(c_1) = 2c_1 - c_1^2$ where $0 \leq c_0 \leq b/a$, $0 \leq c_1 \leq 2$, $w_0 = 0$ and $w_1 = 1$. There is no population growth: $\nu = 1$.

Substituting into (B.9) and rearranging,

$$(B.10) \qquad\qquad c_{0,t+1} = \sqrt{ac_{0,t} - bc_{0,t}^2}.$$

If we specified the parameters of function v differently, then before taking the square root we would receive a quadratic equation containing also a first power term. (B.10) is a nonlinear first-order scalar difference equation. It is noteworthy that we obtain the square root of the famous logistic equation (Example 3.3) playing so important role in chaos theory. As is known, the results obtained for the logistic equation are *universal*, that is, they are qualitatively valid for any unimodal smooth scalar function, including the R.H.S. of (B.10). The system has two steady states: the trivial autarky ($c_{0,B} = 0$ and $c_{1,B} = 1$) and the golden rule state: ($c_{0,G} = a/(1+b)$ and $c_{1,G} = 1 - c_{0,t}$).

Dynamical systems frequently have cycles, too, which return to the initial states after a finite time. Generally it is quite difficult to find cycles with longer periods. However, in this case, the method of Example 3.5 applies, since for certain parameter vectors, there exists a 2-cycle. Let us experiment with $x_1 = c_{0,2k-1}$ and $x_2 = c_{1,2k}$. Taking squares, (B.10) leads to

$$x_2^2 = ax_1 - bx_1^2 \quad \text{and} \quad x_1^2 = ax_2 - bx_2^2.$$

Inserting the second equation into the first, we obtain a fourth degree equation for x_2. Since the steady states also satisfy that equation, we can factor out x_2 and $x_2 - a/(1+b)$. Obtaining a quadratic equation, we have the two roots: $x_{1,2} = a(1 \pm \sqrt{1 - 4/(1+b)})/[2(b-1)]$.

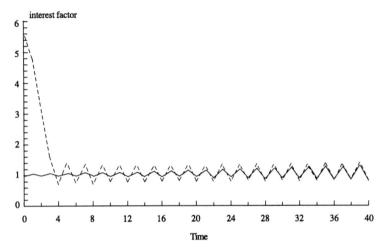

Figure B.2
Dual instability in the time domain: quadratic utility function

(Because of symmetry, the "two" versions of x_2 are just x_1 and x_2.)
The trial succeeded, we have the following 2-cycle:

$$c_{0,2k-1} = x_1; \quad c_{1,2k-1} = 1 - x_1 \quad \text{and} \quad c_{1,2k} = x_2; \quad c_{1,2k} = 1 - x_2.$$

Until now we have studied the generalized version of Gale's example of $a = 5$ and $b = 4$. Gale did not investigate if his 2-cycle is stable or not. It is unstable. We could study the stability conditions of the steady state as well as of the cycle by evaluating $|f'(x_F)|$ and $|f'(x_1)f'(x_2)|$ at the suitable values. Instead of doing so, we turn to our other method.

Example B.6. Generalization of Example B.5. Retain everything in Example B.5 but replace $w_0 = 0$ with $0 < w_0 < 1$. Renouncing the possibility of having a simple recursion (B.10), we return to the generally simpler method of working with interest factor rather than consumption vectors. We have the

Young's conditional consumption and saving functions

$$c_0(r) = \frac{a - w_0 r + w_0 r^2}{b + r^2} \quad \text{and} \quad s(r) = \frac{b w_0 - a + w_0 r}{b + r^2}.$$

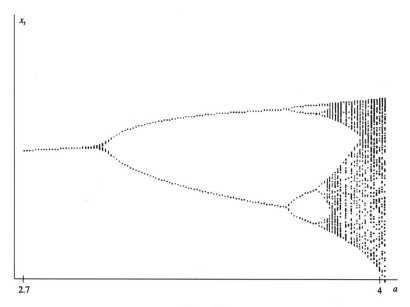

x_t

2.7   4 a

Figure B.3
Bifurcation diagram: OLG

$S(r_t, r_{t+1}) = 0$ yields a quadratic equation on r_{t+1}^2 which may have generally no solution, one solution or more than one solution. Avoiding the detailed calculations of the stability conditions, we only mention the case $a = 4.8$, $b = 6$, $w_0 = 0.4$ ($r_B = 6$), displayed in Figure B.2 with initial states $r_0^1 = 1.04$ and $r_0^2 = 5.96$, respectively. Observe that both steady states are unstable and the emerging 2-cycle is stable. (We have chosen initial states which are not too close to the steady states. Otherwise, the convergence to the 2-cycle would take too much time.)

Having now the tools of chaos theory at our disposal, we can produce stable cycles in a series. With the help of a very short computer program, the modified Problem 3.6 makes a new bifurcation diagram, which reveals the stability of the system for a wide zone of a under the restriction $b = a$. Measure $a \in [2, 7; 4]$ on the horizontal axis, with step-size 0.01; and x_t between $t = 100$ and 200 on the vertical axis, always starting from the initial state $x_0 = 0.6$. It can be seen again that by increasing a, stable steady states are followed by limit cycles followed by chaotic paths. (The comparison with the bifurcation diagram in Figure 3.6 highlights that the square root does not change the qualitative properties of the logistic iteration.) As a completion, here are two problems to be solved.

Problem B.6. a) Prove that for Leontief utility function ($\mu = 1$) and $w_0 \neq w_1$, the corresponding difference equation simplifies into $r_{t+1} = 1/r_t$. b) Observe that each initial state generates a golden rule 2-cycle.

Problem B.7. (Aiyagari, 1989, p. 167. ftn. 4.) Consider a discounted CRRA utility function. a) Specify $S(r_t, r_{t+1}) = 0$ for a golden rule 2-cycle where $r_{t+1} = 1/r_t$. b) Write down the explicit solution for $\mu = 2/3$.

Other models and conclusions

For lack of space, we do not study the *productive OLG economy* where the interest rate and wage are equal to the marginal products of capital and labor, respectively (for example, Diamond, 1965). Already the pioneering article of Samuelson (1958) emphasized that without introducing *social security* or *money*, our OLG economy does not work. The application of these results to the comparison of two social security systems, namely, the unfunded and the funded system (Aaron, 1966) is especially helpful but is also omitted. Blanchard and Fischer (1989) and Azariades (1993) are particularly good references. Balasko and Ghiglino (1995) provides a remarkable existence theorem.

Having arrived at the end of Appendix B, it seems to be useful to summarize the conclusions.

The family of OLG models explains successfully some phenomena of a dynamic economy which are connected with the interactions of fathers and sons (mothers and daughters). Certain theorems of optimal growth theory (for example, golden rule) retain their validity, others (like stability) lose them. As a results of the preserved golden rule, the condition of superiority of unfunded with respect to funded system obtains. It is a novelty, however, that in OLG not all optimal path are stable, there are cycles as well as chaotic paths.

On the other hand, in the family of OLG models it is very restrictive that at any given time, only two generations overlap, (the old and the young). Thus the length of the time-period is very long, say 30 years and within this time-period, no changes can be considered. We only list several unrealistic consequences. (i) The time spent in labor and in retirement are equal. (ii) The earnings do not increase with age. (iii) Even the shortest (2-)cycle, with a period of 60 years, is too long (Sims, 1986). Unfortunately, the experts of OLG often forget about these (and other) limitations and draw practical conclusions with excessive self-confidence.

In Appendix C we shall show how the assumptions and theorems are modified if Overlapping Generations (OLG) models are generalized to Overlapping Cohorts (OLC) models.

APPENDIX C. OVERLAPPING COHORTS

Appendix B analyzed the family of Overlapping Generations models. In this Appendix we shall consider an *Overlapping Cohorts* (for short, OLC) model where many rather than two cohorts are destined to live together by the nonstorability of the goods consumed by them.

Section C.1 considers an *OLC exchange economy* driven by the so-called mixed expectations. Section C.2 deals with stationary paths, or using another name, the existence and uniqueness of the *steady states*. Sections C.3 and C.4 investigate general (nonstationary) paths under *rational and naive expectations*, respectively. This Appendix is based on Molnár and Simonovits (1996) and (1998) but further interesting information can be found on steady states and cycles in Simonovits (1995a) and (1999b), respectively. Appendix B and C demonstrate the difference between an aggregated and a disaggregated system.

C.1 AN EXCHANGE ECONOMY

Overlapping Cohorts

We use the word *cohort* rather than the accepted term *generation*, because in the first meaning of the word no more than three or four generations may live together. Moreover, the expression *two overlapping generations* may hide the fact that overlapping generations models lump together many different cohorts into two generations. A further problem arises that the bulk of the models consider so-called 2-generation models (excluding children), and they also neglect

253

intra-generation heterogeneity (homogeneous generations). The correct solution is obviously as follows: divide the population into as many cohorts as justified by the analysis. Classical examples are Yaari (1965), Tobin (1967), Arthur and McNicoll (1978), Elbers and Weddepohl (1983), Peters (1988) in continuous time and Aaron (1966), Gale (1973), Auerbach and Kotlikoff (1987), Augusztinovics (1989) and (1992) in discrete time. I have already referred to my own and co-authored papers. We shall then speak of Overlapping Cohorts models, and use the abbreviation OLC.

There is an additional source of confusion which should be mentioned at this point: Balasko et al. (1980) proposed a method to transform multicohort one-good models into 2-cohort, multigoods models. However, persons applying this method frequently forget that at the reduction, the number of goods increases appropriately (as was emphasized by Kehoe and Levine (1984, p. 91) and Reichlin (1992)).

We shall see that the 2-generation (more properly, 2-cohort) OLG assumptions are far from being as innocent as their users seem to believe. These assumptions exclude the earning differences between young and older workers, and the consumption by children. (Samuelson (1975) defended the exclusion of children's consumption on the ground that it is relatively negligible but the overpopulation of the Third World undermines this claim.)

We shall determine the interest factors from the model and in addition to steady state, analyze cyclic and other paths. It is a technical simplification that we neglect the growth (modelled in Appendix B) and death risk (treated in Sections 8.1 and 10.1).

Assume that the number of 'newborns' is time-invariant, for simplicity, it is equal to 1. People live exactly for $D + 1$ periods, indexed by $i = 0, 1, \ldots, D$, $D > 1$. Then each date the number of people aged i is equal to 1. Calendar time is denoted by $t = \ldots - 2, -1, 0, 1, 2, \ldots$.

Obviously, $w_i \geq 0$. To have a nonempty model, at least one element of the earning vector should be positive. Let L and M be the first and the last positive components of the earning vector, respectively: $0 \leq L \leq M \leq D$. Figuratively, we shall speak of cohorts of *children, workers* and *pensioners* for $0 \leq i < L$, $L \leq i \leq M$ and $M < i \leq D$, respectively. For notational simplicity, we will normalize the earning vector: $\sum_{i=0}^{D} w_i = 1$.

Let $c_{i,t}$ and $s_{i,t}$ be the consumption and the saving of a person of age i at time t, respectively. $c_{i,t} \geq 0$ and $s_{i,t} = w_i - c_{i,t}$.

We shall make the simplifying assumption that there exists a perfect annuity market where everybody can freely sell his expected stream of earnings to an insurance company and in turn can buy an expected consumption path. Everybody has assets, which – from time to time – can be positive, negative or zero. These assets are managed by the foregoing insurance company which pays or charges interest at every time-period.

Let r_t be the *interest factor* (that is, one plus interest rate) at date t. We shall need the *compounded interest factor* for $(t, t + i]$:

$$(\text{C.1}) \qquad R_{t,t+i} = r_{t+1} \cdots r_{t+i}, \qquad (R_{t,t} = 1).$$

Although in reality a baby cannot borrow, we shall assume that until reaching adulthood, his wealth is managed by his parents. It is assumed that earnings as well as expenditures are accounted at the end of each time-period. Here is the dynamics of *assets of a consumer aged i at the end of time-period t*:

$$(\text{C.2}) \qquad a_{i,t} = r_t a_{i-1,t-1} + w_i - c_{i,t}, \qquad a_{-1,t} = 0;$$

or equivalently,

$$(\text{C.3}) \qquad a_{i,t} = \sum_{j=0}^{i} R_{t-i+j,t} \; s_{j,t-i+j}.$$

To avoid unnecessary repetitions, the domains of i and t are omitted if no confusion arises.

A saving path $\{s_{i,t+i}\}_{i=0}^{D}$ leaves *zero bequest* if $a_{D,t+D} = 0$, or equivalently, the discounted lifetime saving (the present value) is equal to zero:

$$(\text{C.4}) \qquad \sum_{i=0}^{D} R_{t,t+i}^{-1} \; s_{i,t+i} = 0.$$

Let S_t be the total saving of the society at date t:

$$S_t = \sum_{i=0}^{D} s_{i,t}.$$

A saving profile $\{s_{i,t}\}_{i=0}^{D}$ is *feasible* if the total saving of the society is equal to zero:

$$(\text{C.5}) \qquad S_t = 0.$$

We shall assume that consumers leave zero bequest and the feasibility condition is satisfied: (C.4) and (C.5) hold. Note that for stationary zero interest rates ($r_t \equiv 1$) and stationary saving paths ($s_{i,t} \equiv s_{i,0}$ for every i and t), (C.4) is equivalent to (C.5).

Let A_t be the *total assets* of the population at the end of time-period t:

$$A_t = \sum_{j=0}^{D} a_{j,t}.$$

By (C.5) and $a_{D,t} = 0$, $A_t = r_t A_{t-1}$.

Expectations

We recapitulate Section 4.5 in our OLC framework. Expectations will play a crucial role in the analysis. Using the customary notation, $_t r_\tau$ will denote the expected interest factor which the forecaster makes at time-period t for future time-period $\tau(> t)$.

Rational expectations
Each expected interest factor is equal to the *actual* (future) interest factor:

(C.6) $\qquad\qquad _t r_{t+i} = r_{t+i}; \qquad i = 1, \ldots, D.$

Naive expectations
Each expected interest factor is equal to the *current* interest factor:

(C.7) $\qquad\qquad _t r_{t+i} = r_t; \qquad i = 1, \ldots, D.$

Mixed expectations
We make the following generalization of these two types of expectations (see Section 4.5). Let d be an integer: $0 \le d \le D$.

(i) In addition to the current interest factor r_t, the *near* future interest factors $r_{t+1}, \ldots, r_{t+d}$ are known at date t:

$$_t r_{t+i} = r_{t+i}, \qquad i = 1, \ldots, d.$$

(ii) Expectations on *far* future interest factors $r_{t+d+1}, \ldots, r_{t+D}$ are equal to the interest factor $t + d$:

(C.8) $\qquad\qquad _t r_{t+i} = r_{t+d}, \qquad i = d + 1, \ldots, D.$

The set of (i)–(ii) is indeed a generalization, since rational and naive expectations are obtained for $d = D$ or 0, respectively. Note also that for $D = 1$, (i)–(ii) collapse to either rational expectations ($d = 1$) or naive expectations ($d = 0$). For $d = 0$ and $d = n$, (i) and (ii) are empty.

At date t a consumer of age i has the following expected budget constraint:

$$(C.9) \qquad r_t a_{i-1,t-1} + \sum_{j=0}^{D-i} {}_t R_{t,t+j}^{-1}\, {}_t s_{i+j,t+j} = 0.$$

In view of our framework,

$${}_t R_{t,t+j} = R_{t,t+j} \quad (j \le d); \qquad {}_t R_{t,t+j} = R_{t,t+d}\, {}_t R_{t+d,t+j} \quad (j > d).$$

Expected individual consumption at age i depends on the expected interest factors ${}_t r_\tau$ and the past financial asset $a_{i-1,t-1}$:

$$(C.10) \qquad {}_t c_{i,t} = c_i(a_{i-1,t-1}, r_t, \ldots, {}_t r_{t+D-i}),$$

satisfying (C.9).

Dynamics

We introduce notations

$$a_t = (a_{0,t}, \ldots, a_{D-1,t})^{\mathrm{T}}, \quad {}_t r = (r_t, \ldots, r_{t+d-1})^{\mathrm{T}}, \quad x_t = (a_t,\ {}_t r)^{\mathrm{T}}.$$

(If $d = 0$, then ${}_t r$ is dropped as empty.) Under suitable assumptions, at date t (C.8) determines the new forecasts ${}_t r_{t+d+1}, \ldots, {}_t r_D$, (C.10) decides the new consumption forecasts, while (C.5) and (C.9) jointly determine the critical interest factor r_{t+d}. Thus a vector difference equation $x_t = f(x_{t-1})$ is defined. For $d > 0$, future interest rates $r_{t+1}, \ldots, r_{t+d-1}$ should be known in advance. We shall see when and how the *indeterminacy* problem can be alleviated. The system started its functioning at time-period 0, and $x_{-1} = (a_{-1,-1}\, r)^{\mathrm{T}}$ is called the *vector of initial conditions*.

Time-invariant solutions play an important role in any dynamic analysis. A $(D + d)$-dimensional vector x_F is called a *steady state* if it is mapped into itself: $x_F = f(x_F)$. For short, we shall often refer to a steady state as a_F or r_F. The meaning of A_F is also obvious.

Following Gale (1973, Part II), we make now a few observations. At a steady state, $A_F = r_F A_F$ holds, suggesting the following distinction. If r_F differs from 1, it is referred to as *balanced* (or *nonmonetary*) steady state and will be denoted by r_B, ($A_B = 0$). If $r_F = 1$, then the feasible steady state is called *golden rule* steady state and will be denoted by A_G ($r_G = 1$). In this latter case we shall distinguish three subcases: a) if $A_G < 0$, then the steady state is *debtor* (classical); b) if $A_G > 0$, then the steady state is *creditor* (Samuelson, or monetary); c) if $A_G = 0$, then the steady state is *symmetric* (coincidental). As can be shown, these definitions are consistent with Appendix B.

Before introducing optimization, we present an elementary theorem which can be traced back to Samuelson (1958) and Gale (1973, p. 33) for rational expectations and OLG framework.

Theorem C.1. *Mixed expectations.* a) *If a golden rule steady state r_G is (asymptotically) stable, then the initial conditions satisfying $A_{-1} \leq 0$ and $A_{-1} \geq 0$ yield unstable paths for $A_G > 0$ and $A_G < 0$, respectively.*

b) *If a balanced steady state r_B is stable, then $r_B \leq 1$.*

Remark. Note that global stability is excluded.

Proof. Consider $A_t = r_t A_{t-1} = \cdots = R_{-1,t} A_{-1}$.

a) $A_{-1} \leq 0$ implies $A_t \leq 0$, thus A_t cannot converge to $A_G > 0$. The proof is identical for $A_{-1} \geq 0$.

b) If the steady state r_B is stable and contrary to the theorem, $r_B > 1$, then for a suitable positive constant α, $R_{-1,t} \geq \alpha r_B^{t+1}$, that is, for any nonzero A_{-1}, $|A_t| \geq \alpha r_B^{t+1} |A_{-1}|$, that is, A_t is divergent, a contradiction. ∎

Optimization

Conforming to the tradition of neoclassical economics, from now on we shall derive the consumption and saving decisions from optimization of the agents of different ages.

Let $U(c_0, \ldots, c_D)$ be the *timeless* utility function of a representative agent, which is concave, nondecreasing in all variables and if all variables are increasing, then U is also increasing (local insatiability).

As is usual, we shall work with *discounted* utility functions of the form

(C.11) $$U(c_0, \ldots, c_D) = \sum_{i=0}^{D} \beta^i u(c_i), \qquad 0 < \beta \leq 1$$

where $u(c)$ is a *time-period utility (felicity) function*, strictly concave and increasing; β is the *discount factor*.

For a given expected interest factor path, the consumer born at time $t - i$ maximizes his (expected) remaining utility function $U_i({}_t c_{i,t}, \ldots, {}_t c_{D,t+D-i})$ under the condition of zero (expected) bequest (C.9). There exists a unique *conditionally optimal* consumption path.

But at any time-period we have $D + 1$ cohorts with (C.5) constraining their collective behavior. We shall call an interest factor path derived from optimization and satisfying (C.5) *competitive*. It is to be seen if for proper initial conditions, the dynamics of interest factors is defined or not. Note that different types of expectations yield different paths.

Constant relative risk aversion (CRRA) utility functions are quite general and they play an important role in the analysis of life-cycle problems. Let σ be a real number, $-\infty < \sigma < 1$, $\zeta = 1 - \sigma$ is referred to as the *coefficient of relative risk aversion*. Then the time-period utility function is given by

(C.12) $$u(c) = \begin{cases} \sigma^{-1} c^{\sigma} & \text{if } \sigma \neq 0; \\ \log c; & \text{if } \sigma = 0. \end{cases}$$

(C.11) and (C.12) imply the total utility function

(C.13) $$U(c_0, \ldots, c_D) = \begin{cases} \sigma^{-1} \sum_{i=0}^{D} \beta^i c_i^{\sigma} & \text{if } \sigma \neq 0; \\ \sum_{i=0}^{D} \beta^i \log c_i & \text{if } \sigma = 0. \end{cases}$$

We shall frequently use the following transformation of σ: $\mu = \sigma/(\sigma - 1)$ where $1 - \mu$ is the *intertemporal elasticity of substitution*. (Note that CRRA and CES only coincide with time-separable utility functions.)

As is known, after suitable transformations, Cobb–Douglas and Leontief utility functions are obtained for $\sigma = 0$ ($\mu = 0$) and $\sigma = -\infty$ ($\mu = 1$), respectively. While the former is very popular, the latter is hardly used (notable exception are Ando and Modigliani, 1963, Assumption IV and Augusztinovics, 1992).

Although the CRRA utility functions are concave for the entire parameter interval $\sigma \in [-\infty, 1)$, we shall assume that $\sigma \leq 0$, that is, the risk aversion coefficient is equal to or larger than 1. Interval $\sigma \in (0, 1)$ is omitted, that is, we shall exclude $\mu < 0$. In the dismissed case a consumption path with some zero components would not give low enough utility to make it infeasible (see for example, Laitner, 1984, p. 114).

Competitive consumption path

The full description consists of the following elements (for $d = D = 1$, see Kehoe and Levin, 1990). At the end of time-period $t - 1$ the interest factors $r_t, \ldots, r_{t+d}$ and consumer of age i's financial asset $a_{i-1,t-1}$ are given (r_{t+d} is to be determined later). Consumer of age i starts his conditional optimization on his remaining expected consumption paths $_tc_{i,t}, \ldots, _t c_{D,t+D-i}$. To do so the following notations are needed:

Truncated utility function at age i

$$(C.14) \qquad U_i\left(_tc_{i,t}, \ldots, _tc_{D,t-i+D}\right) = \sigma^{-1} \sum_{i=0}^{D-i} \beta^j \, _tc_{i+j,t+j}^{\sigma}.$$

Expected present value of remaining life-time earnings at t

$$(C.15) \qquad W_{i,t} = \sum_{i=0}^{D-i} w_{i+j} \, _tR_{t,t+j}^{-1}.$$

Expected present value of remaining life-time corrected discount

$$(C.16) \qquad V_{i,t} = \sum_{i=0}^{D-i} \Phi^j \, _tR_{t,t+j}^{-\mu} \quad \text{where} \quad \Phi = \beta^{1-\mu}.$$

Theorem C.2. *Mixed expectations. For a CRRA utility function (C.14), subject to (C.9), the conditionally optimal (generally infeasible) consumption at age i in time t is given by*

$$(C.17) \qquad c_{i,t} = \frac{r_t a_{i-1,t-1} + W_{i,t}}{V_{i,t}}.$$

Problem C.1. Prove Theorem C.2.

Remark. Note that for $\sigma = -\infty$, $\mu = 1$ and $\Phi = 1$ hold and (C.16)–(C.17) are well-defined. The case of $\sigma = 0$ ($\mu = 0$) is similar.

Having determined the conditional optima, we turn to the feasibility condition. Substituting (C.17) into the definition of S_t and (C.5) yields

$$\text{(C.18)} \qquad S_t = 1 - \sum_{i=0}^{D} \frac{r_t a_{i-1,t-1} + W_{i,t}}{V_{i,t}},$$

$$\text{(C.19)} \qquad \sum_{i=0}^{D} \frac{r_t a_{i-1,t-1} + W_{i,t}}{V_{i,t}} = 1.$$

We can announce

Theorem C.3. *Mixed expectations. Under CRRA utility functions (C.14), a competitive interest factor r_{t+d} (if exists at all) satisfies (C.15)–(C.16) and (C.19).*

Simplified rational expectations

To avoid the difficulties of optimization for mixed expectations, here we specify them to rational expectations [(C.6)], moreover, apply Gale's (1973) *short-cut*. Assume that the system started at date $-\infty$ rather than at 0, and it behaved optimally since the beginning. Then we have

Theorem C.4. *Short-cut rational expectations. The optimal (generally infeasible) consumption plan at age i in time t is given by*

$$\text{(C.20)} \qquad {}_t c_{j,t} = \Phi^j R_{t-j,t}^{1-\mu} H_{t-j}$$

where

$$\text{(C.21)} \qquad H_t = \frac{W_{t,0}}{V_{t,0}}.$$

Proof. Under rational expectations, there is no reason to recalculate the optimal paths at every occasion. (C.17) reduces to (C.20)–(C.21) for $i = 0$. Note that $c_{0,t} = H_t$. ∎

Having determined the conditional optima, we check now the feasibility condition.

The complex approach would yield a $2D$-dimensional mixed system $x_t = f(x_{t-1})$, but under the short-cut it is replaced by a $(2D-1)$-dimensional pure system

$$r_{t+D} = g\left(r_{t-D+1}, \ldots, r_{t+D-1}\right).$$

The latter is much easier to study as well as to display than the former. From now on we generally omit the adjectives *complex* and *short-cut*.

Substituting the conditional optimum into (C.18), yields the *total saving function* as a function of the relevant interest factors:

(C.22)
$$S_t = 1 - \sum_{i=0}^{D} \Phi^i R_{t-j,t}^{1-\mu} H_{t-i}.$$

With the aid of (C.22) we can calculate function g. The analysis of the implicit difference equation $S_t = 0$ is generally quite involved, therefore we postpone it to Section C.3. As a preparation, we shall consider the steady state in Section C.2.

C.2 STEADY STATES

Although the steady states are independent of the value of d in mixed expectations, we postponed the investigation of the steady states after spelling out the simplified model of rational expectations.

Existence and uniqueness

We also recapitulate Kim's (1983) condition on the *regularity of earnings*: the worker starts working before the last time-period [D] and does not retire before the second time-period [1]. In formula: $L < D$ and $M > 0$.

For simplified rational expectations, we need only the first members rather than the entire sequence $\{W_{i,t}, V_{i,t}\}$. Therefore it will be useful to introduce the short-hand $W_t = W_{0,t}$ and $V_t = V_{0,t}$. Formula (C.20) and its auxiliary formulas can be reduced with the aid of the timeless interest factor r:

(C.15°)
$$W(r) = \sum_{i=0}^{D} w_i r^{-i},$$

$$(C.16^\circ) \qquad V(r) = \sum_{i=0}^{D} \Phi^i r^{-\mu i},$$

$$(C.21^\circ) \qquad H(r) = \frac{W(r)}{V(r)},$$

$$(C.20^\circ) \qquad c_j(r) = \Phi^j r^{(1-\mu)j} H(r).$$

Substituting (C.15°)–(C.16°) and (C.20°)–(C.21°) into (C.22), yields the macro saving-function:

$$(C.23) \qquad S(r) = 1 - \sum_{i=0}^{D} \Phi^j r^{(1-\mu)j} H(r).$$

Following Kim, we extend the validity of Theorem 7 in Gale (1973) from weak to strong risk aversion, but confining attention the CRRA utility functions. We shall need the following notations:

$$(C.24) \quad \mu_1 = \min\left(\frac{M}{D}, 1 - \frac{L}{D}\right) \quad \text{and} \quad \mu_2 = \max\left(\frac{M}{D}, 1 - \frac{L}{D}\right).$$

Obviously, the roots of $S(r) = 0$ are the interest factors belonging to the steady states. We present the following theorem without proof.

Theorem C.5. *(Kim, 1983 and Simonovits, 1995a, Theorem 9.) Let the earning path be regular, the utility function be CRRA with strong risk aversion ($\mu \geq 0$):*
a) If $0 \leq \mu < \mu_1$ holds, then there exists at least one balanced steady state $r_B > 1$ for the debtor case, $r_B < 1$ for the creditor case, and $r_B = 1$ for the symmetric case.
b) If $\mu_2 < \mu < 1$ holds, then there exists a balanced steady state $r_B < 1$ for the debtor case, $r_B > 1$ for the creditor case, and $r_B = 1$ for the symmetric case.
c) If $\mu_1 < \mu < \mu_2$ (window) holds, then there exists either no balanced steady state or there exists more than one balanced steady state.

Remarks. 1. Note that μ_1 and μ_2 are defined respectively as the minimum and the maximum of the following two ratios of three time spans: (childhood + working life)/life and (working life + retirement)/life. As Kim (1983) emphasized, the literature had only considered case a) where the creditor case is not Pareto-optimal: $r_B < 1$.

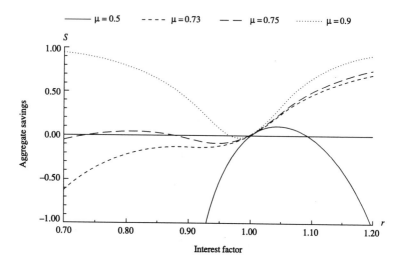

Figure C.1
Interest factor and savings

But case b) can be as relevant as case a), and then it is the debtor case which is not Pareto-optimal: $r_B < 1$.

2. In Theorem B.1 earnings in both time-periods were positive and autarky was the only balanced steady state.

In a realistic setting the natural physical length of a time-period is one year, thus the number of overlapping cohorts is about one hundred. Since we do not take into account death risk, we shall generally work with $D = 71$.

To illustrate the situation, function $S(r)$ is displayed for four different values of μ. Figure C.1 presents one case, namely, $L = 20$, $M = 57$ and $\beta = 0.99$.

Problem C.2. Using computer, demonstrate that for $\mu = 0.73$, there is no balanced steady state, while for $\mu = 0.75$, there are at least two: $r_{B_1} = 0.7375$ and $r_{B_2} = 0.8691$. (It is an open question if there can be two balanced steady states around 1.)

To illustrate the concepts, Table C.1 presents the dependence of the balanced steady states on the key parameters of the model. We have fixed $D = 71$, $m = 18$, $M = 55$, combined $\mu = 0.5$, 0.75 and 1 with $\beta = 0.98, 0.99$ and 1 to obtain 7 realistic values of the parameter vector (μ, β). (The last column will only be needed at the end of this Section.) For simplicity, we assumed flat earning profiles in all the numerical calculations apart from Theorem C.8: $w_i = 1/(M - L + 1)$ if $i = L, \ldots, M$ and $w_i = 0$ otherwise (see Ando and Modigliani, Assumption IV).

Table C.1. *Characteristics of balanced steady states*
($L = 18$, $M = 55$, $D = 71$)

Anti-elasticity	Discount factor	Balanced interest factor	Zero of denom-inator
μ	β	r_B	$r_D - r_B$
0.50	0.98	1.089795	-0.000043
0.50	0.99	1.053719	-0.000072
0.50	1.00	1.016825	-0.000190
0.75	*0.98*	*0.928033*	*-0.000016*
0.75	0.99	0.954918	-0.000043
0.75	1.00	0.978895	-0.000128
1.00		0.993590	-0.000150

Remark. The row in italics represents Examples C.4 and C.8.

In certain analytical investigations, however, we do not distinguish that many cohorts and work with less cohorts. To evaluate the resulting magnitudes correctly, we should rely on the principle of *scale-invariance*: if people live $T + 1$ years and we distinguish $D + 1$ cohorts, then the cohort interest factors should be expressed as *annual interest factors*: $r = \mathbf{r}^{(T+1)/(D+1)}$. A similar relation holds for discount factors: $\beta = \beta(T)^{(T+1)/(D+1)}$.

The next problem demonstrates the peculiar role of $r_B = 1/\beta$ in the life-cycle models.

Problem C.3. Flat CW profiles without children and pensioners. Prove that $w_i = 1/(D + 1)$, $i = 0, \ldots, D$, implies $r_B = 1/\beta$ and $c_j = 1/(D + 1)$, $j = 0, \ldots, D$ (autarky).

Here is another example for an autarch balanced state which at the same time, is also golden rule state.

Example C.1. Golden rule autarky. Let us assume that the earning sequence is a geometric sequence given by the corrected discount factor as a quotient:

$$w_j = \frac{\Phi^j}{\sum_{i=0}^{D} \Phi^i}, \quad j = 0, 1, \ldots, D.$$

Then by (C.20°), $c_j(r) = w_j r^{(1-\mu)j}$, $j = 0, 1, \ldots, D$. Here the individual as well as the total saving is positive for $r > 1$ and negative for $r < 1$, that is, the autarch steady state is realized, that is, $r_B = 1$, regardless of D (see Ghiglino and Tvede, 1995).

C.3 DYNAMICS WITH RATIONAL EXPECTATIONS

Dynamics

Having finished our long preparation, we turn to the analysis of genuine dynamics. In this Section we retain the assumption of rational expectations: (C.6), in the simplified form discussed at the end of Section C.2. We shall express r_{t+D} as an implicit or explicit function of $r_{t-D+1}, \ldots, r_{t+D-1}$. To do so, we separate the terms containing r_{t+D} in W_t, V_t and S_t. The remaining sums are distinguished from their counterparts by *bold* symbols:

$$(C.25) \qquad \mathbf{W}_t = \sum_{i=0}^{D-1} w_i R_{t,t+i}^{-1},$$

$$(C.26) \qquad \mathbf{V}_t = \sum_{i=0}^{D-1} \Phi^i R_{t,t+i}^{-\mu},$$

$$(C.27) \qquad \mathbf{S}_t = 1 - \sum_{j=1}^{D} \Phi^j R_{t-j,t}^{1-\mu} H_{t-j}.$$

Using the feasibility condition (C.19), with simple calculations, the following equation is obtained:

$$\mathbf{S}_t = H_t = \frac{\mathbf{W}_t + w_D R_{t,t+D-1}^{-1} r_{t+D}^{-1}}{\mathbf{V}_t + \Phi^D R_{t,t+D-1}^{-\mu} r_{t+D}^{-\mu}}$$

which yields

Theorem C.6. *Rational expectations.* (i) *Assuming it exists, the competitive interest factor path is determined by the following implicit difference equation of order $2D - 1$:*

(C.28) $w_D R_{t,t+D-1}^{-1} r_{t+D}^{-1} - \mathbf{S}_t \Phi^D R_{t,t+D-1}^{-\mu} r_{t+D}^{-\mu} - \mathbf{S}_t \mathbf{V}_t + \mathbf{W}_t = 0.$

(ii) *If $0 < \mu \leq 1$, $w_D = 0$ and the function in (C.29) below is defined, then (C.28) reduces to the explicit difference equation of order $2D - 1$:*

(C.29) $$r_{t+D} = \left(\frac{\mathbf{S}_t \Phi^D}{\mathbf{W}_t - \mathbf{S}_t \mathbf{V}_t} \right)^{1/\mu} \frac{1}{R_{t,t+D-1}}.$$

Remarks.   1. As was the case in the general model of mixed expectations in Section 4.5, it is the OLG feasibility condition in time-period t that determines the optimal interest factor of $t+D$. In annual account and normal lifespan this means that the current interest factor was determined several decades ago. This shows the lack of real-time dynamics with rational expectations in our case. (Kehoe and Levin (1990) speaks of *indeterminacy*.)

2. At the same time, it is just this indeterminacy of rational expectations, which Laitner (1981) and (1984) use to eliminate instability. At the same time, this solution requires such a precision in calculation which can hardly be expected from an ordinary person (see also Kehoe, 1991).

3. This method generalizes the saddle-point method of Section 10.2. The higher the number of the independent variables, however, the weaker the numerical stability of the procedure. Let us remind the reader to Example 1.10 (taken from Example 10.2 in Ralston, 1965): the errors distract the path from the stable arm.

4. The implicit function theorem guarantees the existence of r_{t+D} if

$$\frac{\partial S}{\partial r_{t+D}} = \frac{w_D R_{t,t+D-1}^{-1} r_{t+D}^{-2} V_t - \mu W_t \Phi^D R_{t,t+D-1}^{-\mu} r_{t+D}^{-\mu-1}}{V_t^2} \neq 0.$$

If $r_t \equiv 1$ and $\beta = 1$, then $\dfrac{\partial S}{\partial r_{t+D}} = \dfrac{w_D(D+1) - \mu}{(D+1)^2}$. For realistic models, $w_D(D+1)$ lies between 0 and 1, thus there exists a singular value $\mu_o = w_D(D+1)$ for which $\dfrac{\partial S}{\partial r_{t+D}} = 0$. Note also that $\dfrac{\partial S}{\partial r_{t+D}}$

is a decreasing linear function of μ, its coefficient is $-1/(D+1)^2$. The order of magnitude of $\dfrac{\partial S}{\partial r_{t+D}}$ is also $1/(D+1)^2$. Turning to the principle of scale-invariance, we express the partial derivative in annual terms (in bold):

$$\frac{\partial S}{\partial \mathbf{r}_{t+D}} = \frac{\partial S}{\partial r_{t+D}} \frac{dr_{t+D}}{d\mathbf{r}_{t+D}} \approx \frac{T+1}{(D+1)^3}.$$

For fine enough resolution (years or months), this quantity is hardly different from 0. Thus at least around the golden rule steady state of the undiscounted case, the implicit function (C.28) is *ill-conditioned*. In contrast, in the popular case, studied in Section B.1, $D = 1$, $w_1 = 0$, only the Cobb–Douglas case is singular, the coefficient $\dfrac{\partial S}{\partial r_{t+1}} = -\mu/4$ is quite large, especially on an annual basis: $\dfrac{\partial S}{\partial \mathbf{r}_{t+1}} = -9\mu$.

Let us remind to the remark made in Example B.6: the solution cannot be continued if the implicit function [(C.28)] has no solution, or what is equivalent, if the R.H.S. of (C.29) is not defined. Then the system becomes unviable. Of course, there are no such problems with the steady state paths, at least not in theory.

Perhaps Theorem C.1 led to Gale's (1973, p. 12 and p. 16) *claim* for $D > 1$, which can be reformulated as follows: *under rational expectations, the lower steady state is stable.* Contrary to Gale's claim (see also the Remark to Theorem B.4), there is no reason to expect so simple result to be true. (The 3-cohort Problem C.4 below is a counter-example to the most cautious formulation of the claim.)

Till the end of this Section, we shall assume that there are child and pensioner cohorts without earnings: $w_0 = \cdots = w_{L-1} = w_{M+1} = \cdots = w_D = 0$, that is, $L > 0$ and $M < D$, thus we can study (C.29). To simplify notations, we shall only write $w_0 = w_D = 0$. For $\mu = 0$, (C.29) now becomes meaningless, therefore we assume $\mu > 0$.

Local instability

We shall continue the discussion with local analysis around the steady state. Following the method outlined in Chapter 3, we linearize $r_{t+D} = g(r_{t-D+1}, \ldots, r_{t+D-1})$. Let $\gamma_{i,\mathrm{F}}$ be the partial derivative of g with respect to r_{t+i}, $i = -D+1, \ldots, D-1$ at point r_{F}.

Then the local g-dynamics around r_{F} is described by

$$r_{t+D}^{\mathrm{d}} = \sum_{i=-D+1}^{D-1} \gamma_{i,\mathrm{F}} r_{t+i}^{\mathrm{d}} \quad \text{where} \quad r_t^{\mathrm{d}} = r_t - r_{\mathrm{F}}.$$

Let $p(\lambda) = \sum_{i=-D+1}^{D-1} \gamma_{i,\mathrm{F}} \lambda^{D-1+i}$ and $P(\lambda) = \lambda^{2D-1} - p(\lambda)$, then $P(\lambda)$ is the characteristic polynomial of the local dynamics g. Let ρ be the maximal modulus of the roots of $P(\lambda)$. By Theorem 3.3, stability only depends on the location of the characteristic roots: all are inside the unit disk (stability) or there are roots outside (instability).

The dynamics of (C.29) may be very complex, even around a steady state it is determined by the roots of a $(2D - 1)$-degree polynomial and $2D - 1$ initial conditions. At the present analysis we frequently had to rely on computer simulation.

Of course, the determination of all $\gamma_{i,\mathrm{F}}$'s would be very difficult and the calculation of ρ may be even harder. We shall use a short-cut and determine only a single coefficient, $\gamma_{-D+1,\mathrm{F}}$. The following Lemma provides a very useful formula for $\gamma_{-D+1,\mathrm{F}}$.

Lemma C.1. *Rational expectations. If $w_0 = 0 = w_D$ and r_{F} is a feasible steady state, then*

$$(C.30) \qquad\qquad \gamma_{-D+1,\mathrm{F}} = r_{\mathrm{F}}^D.$$

Proof. The basic idea is simple: Take $t = 0$ and determine the partial derivative of r_D with respect to r_{-D+1}. Because of $w_D = 0$, $W_t = \mathbf{W}_t$. Note that r_{-D+1} does not appear in either W_0 or V_0, or $R_{0,D-1}^{-1}$. From now on we shall drop the time index 0 of the foregoing functions. Under $r_{-D+2} = \cdots = r_{+D-1} = r_{\mathrm{F}}$, all the functions become one-variable functions of $r = r_{D+1}$. The R.H.S. of (C.29) simplifies into

$$(C.31) \qquad\qquad G(r) = Q(r)^{1/\mu} r_{\mathrm{F}}^{1-D}$$

where

$$(C.32) \qquad\qquad Q(r) = \frac{\mathbf{S}(r)\Phi^D}{W - \mathbf{S}(r)\mathbf{V}}.$$

Denoting $v(r) = W - \mathbf{S}(r)\mathbf{V}$ and subindex $_{\mathrm{F}}$ for values taken at r_{F}, (C.31)–(C.32) imply

$$(C.33) \qquad\qquad G_{\mathrm{F}}' = \mu^{-1} Q_{\mathrm{F}}^{1/\mu-1} Q_{\mathrm{F}}' r_{\mathrm{F}}^{1-D}$$

and

(C.34)
$$Q'_F = \Phi^D \frac{S'_F}{v_F^2}.$$

By the definition of steady state, $r_F = Q_F^{1/\mu} r_F^{1-D}$, that is, $Q_F^{1/\mu} = r_F^D$, hence $Q_F^{1/\mu-1} = r_F^{(1-\mu)D}$. Note that $Q_F = \Phi^D S_F / v_F$ implies $v_F = \Phi^D S_F / r_F^{\mu D}$. Before presenting Theorem C.6, we have already shown that $S_F = H_F$. Thus (C.34) and (C.33) reduce to

$$Q'_F = \frac{W_F r_F^{2\mu D}}{\Phi^D H_F^2} S'_F$$

and

(C.35)
$$G'_F = \mu^{-1} r_F^{\mu D+1} \Phi^{-D} W_F H_F^{-2} S'_F.$$

We have to determine S'_F. Note that $r = r_{-D+1}$ only appears in the last term of S: $\Phi^D r^{1-\mu} r_F^{(1-\mu)(D-1)} H_{-D}(r)$. Dropping now the time index $-D$, thus

(C.36)
$$S'_F = -\Phi^D(1-\mu)r_F^{-\mu} r_F^{(1-\mu)(D-1)} H_F - \Phi^D r_F^{(1-\mu)D} H'_F$$

where

(C.37)
$$H'_F = \frac{W'_F V_F - W_F V'_F}{V_F^2},$$

$$W'_F = -\sum_{i=1}^{D} w_i r_F^{-2} r_F^{-i+1} = -r_F^{-1} W_F$$

(because of $w_0 = 0$),

$$V'_F = -\mu \sum_{i=1}^{D} \Phi^i r_F^{-\mu-1} r_F^{(1-i)\mu} = -\mu r_F^{-1}(V_F - 1).$$

After substitutions, the numerator of H'_F reduces to $r_F^{-1}(\mu-1)W_F V_F - r_F^{-1} \mu W_F$. Using $H = W/V$, (C.37) becomes

(C.38)
$$H'_F = r_F^{-1}(\mu-1)H_F - \frac{r_F^{-1} \mu H_F}{V_F}.$$

Before inserting (C.38) into (C.36), note that both exponents of r_F in (C.36) are equal to $D - 1 - \mu D$.

(C.39) $S'_F = -\Phi^D r_F^{(1-\mu)D-1}[(1-\mu)H_F + H'_F] = \mu \Phi^D r_F^{(1-\mu)D-1} \dfrac{W_F}{V_F^2}.$

Substituting (C.39) into (C.35) yields (C.30).   ∎

Lemma C.1 implies a variant of Theorem 4.11a.

Theorem C.7. *Rational expectations. If $w_0 = 0 = w_D$ and the modulus of at least one root of $P(\lambda)$ is different from 1, then the golden rule steady state r_G is saddle-point unstable.*

Proof. The constant coefficient $\gamma_{-D+1,F} = 1$ is equal to the product of the characteristic roots. We have explicitly excluded the degenerate case of all roots having moduli 1, thus at least one root has a modulus greater than unity (and another has not). ∎

If a balanced steady state r_B is larger than 1, then the trivial Theorem C.1b implies instability. Even if r_B is less than 1, in realistic cases it is close to 1, that is, instability is probable. The next examples illustrate the complications.

Example C.2. Rational expectations, 3-cohorts, CRRA. $L = 1 = M$, $D = 2$, $w_1 = 1$, $0 < \mu \leq 1$. It is easy to show that $r_B = \beta^{(2-2\mu)/(2\mu-1)}$. Hence $r_B < 1$ if $\mu > 1/2$ (debtor); $r_B > 1$ if $\mu < 1/2$ (creditor). For $\mu = 1/2$, there does not exist any balanced root, except in the undiscounted case. The calculations are too long, but one can check with a computer that for $\mu < 1/2$, the golden rule steady state is also unstable. In turn, for $1/2 < \mu < 1$, the balanced steady state is stable. Moreover, for $\mu = 1$, the improbable case occurs: $P(\lambda) = \lambda^3 - 1$, that is, all the moduli are equal to 1, therefore Theorem C.7 does not apply. Computer simulation shows that in this case the system is unstable, but the divergence is very slow. Note that if there is no discount, then this example is closely related to Theorem 4.12a.

Example C.3. Rational expectations, 4-cohort, Leontief utility. $D = 3$, $L = 1$, $M = 2$, $0 < w_1 < 1/2$, $w_2 = 1 - w_1$, $\mu = 1$. It can be shown that $r_B < 1$. Nevertheless, r_B is also unstable, although viable.

At this point we shall introduce a convenient method to analyze high-order systems. Instead of studying function $g : \mathbf{R}^{2D-1} \to \mathbf{R}$, we shall consider its *diagonal restriction* $\Gamma : \mathbf{R} \to \mathbf{R}$ with $r_{-2D+1} = \cdots = r_{-1}$: that is, the *aggregator function* is $r_0 = \Gamma(r_{-1}) = g(r_{-1}, \ldots, r_{-1})$. Obviously, for any steady state r_F, $r_F = \Gamma(r_F)$.

Our computer simulations show that the instability is very strong. Even the *dynamic* verification, that the golden rule and balanced paths are steady states, is difficult for $D > 3$ for generic (debtor or creditor) cases and for $D > 20$ for symmetric cases. In these circumstances we shall speak of *distortion*. We shall now present the *leading example*.

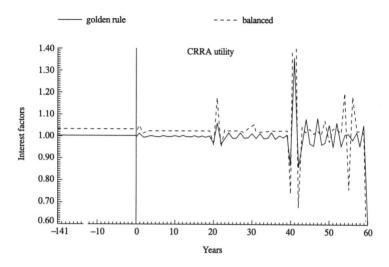

Figure C.2
Instability for rational expectations

Example C.4. Rational expectations, CRRA utility. $D = 71$, $M = 55$, $\mu = 0.75$ and $\beta = 0.98$: $r_B = 0.928033$. (To enhance the quality of presentation, in Figure C.2, 20 rather than 141 initial values are shown for rational expectations.)

Viability

Until now we have only considered local stability around the steady states. We proved and conjectured instability but we have not presented any other information on the dynamics. In a lot of nonlinear systems instability does not mean that the paths explode. Therefore, in principle, it is possible that our system behaves nicely, just oscillates around its unstable steady state without losing its viability, that is, the interest factors remain positive for all time-periods as was seen in Appendix B with cycles. Simonovits (1999b), however, demonstrated that there are no 2-cycles in realistic OLC models, see also Aiyagari (1988), (1989) and Reichlin (1992) on 2-cycles in unrealistic OLC models. In the forthcoming part, we shall argue that the dynamics in realistic overlapping cohorts models with rational expectations is probably not only unstable but unviable as well.

Considering the steady states, we conjecture that a small absolute error in the numerator or the denominator of (C.29) distorts the fraction significantly. In case of the golden rule steady state, the fraction is equal to 1, as it should be, but the denominator (as well as the numerator) is very small for realistic values. While $W(r)$ changes normally with r and stays around 1, $\mathbf{S}_t \mathbf{V}_t$ changes very fast, being the product of two expressions of the order of magnitudes $1/(D+1)$ and D, respectively. The denominator easily becomes negative, while the numerator stays positive.

To those who prefer analytical methods to simulations, we present a special example and a special problem concerning inviability.

Example C.5. Continuation of Example C.3. Rational expectations. Small denominator, 3-cohort, Leontief. $L = 1 = M$, $D = 2$, $w_1 = 1$ and $\mu = 1$. Introducing notation $\xi = 1/r$, we have $W = \xi$, $\mathbf{V} = 1+\xi$, $V = 1+\xi+\xi^2$, $H = \xi/(1+\xi+\xi^2)$, $S = (1-\xi+\xi^2)/(1+\xi+\xi^2)$. Multiplying both the numerator and the denominator by $1 + \xi + \xi^2$, $Q = (1 - \xi + \xi^2)/(-1 + \xi + \xi^2)$ is obtained. We are looking for the zeros of the denominator. Solving the quadratic equation, $\xi_D = (-1 + \sqrt{5})/2$, that is, $r_D = (1 + \sqrt{5})/2 \approx 1.618$. At first sight, this number may appear as quite distant from 1, however, turning to the annual basis, taking its 24-th root, we end up with 1.02. Hence in a 3-cohort model rational expectations are not defined at the quite moderate annual 2 per cent interest rate.

Problem C.4. At this point, we return to the undiscounted Cobb–Douglas utility: $\beta = 1$ and $\mu = 0$, to the positive earnings: $D = 2$, $w_i \equiv 1/3$. Prove that the denominator of (C.29) is equal to $v(r) = -(2/3)r^{-1} + 2 - 2r/3 - r^2/3$, and check with a pocket calculator or a computer that it has a root at 1.355 (or at the annual basis, 1.0135).

Returning to the 7 parameter vectors (μ, β) of Table C.1 of the multicohort case, we calculated the corresponding roots of the denominators around the golden rule. As Table C.2 corroborates, the denominators become zero and then negative for interest factors very close to 1. (The last but one column will only be needed at the end of Section C.4.)

Table C.2. *Characteristics of golden rule steady states*
$$(L = 18, M = 55, D = 71)$$

Anti-elasticity μ	Discount factor β	Spectral radius $\rho(K_G)$	Zeros of denominator $r_D - r_G$
0.50	0.98	0.9699	0
0.50	0.99	0.9752	0.0001
0.50	1.00	0.9876	0
0.75	*0.98*	*0.9931***	*0.0001*
0.75	0.99	1.0334*	0.0001
0.75	1.00	1.0361	0
1.00		1.0073	0.0002

Remarks. 1. Numbers are rounded of four digits. 2. One and two asterisks denote a complex dominant root with an *unstable* and *stable* balanced steady state under naive expectations (Section C.4 below), respectively. 3. The row in italics represents Examples C.4 and C.8.

As the last column of Table C.1 above attests, there are zeros very close to the balanced steady states as well. That singularity appears already for a 3-cohort case, was *analytically* shown in the 3-cohort, Leontief-type Example 6 in Molnár and Simonovits (1996). With such denominators, it is no wonder that the steady states appear as highly unstable and unviable.

Conjecture C.1. *Rational expectations. In any realistic OLC model, paths around the golden rule and the balanced steady states are unviable except for steady states and cycles which are also unstable.*

One could expect that the indeterminacy and instability undermines or at least weakens the popularity of the concept of rational expectations. This is far from the truth. The representatives of the nowadays dominant economics insist on rational expectations, because they are the only expectations which are consistent with the equilibrium thinking. Furthermore, for balanced saddle-points, instability and indeterminacy can be eliminated simultaneously (Section B.5). Since I find this reasoning weak, I am looking for alternatives.

C.4 DYNAMICS WITH NAIVE EXPECTATIONS

In this Section we shall study naive expectations: (C.7) and discuss the corresponding dynamics.

In Section C.3 we have assumed that the consumers have perfect foresight. Having experienced difficulties concerning stability and viability, we resort to the much criticized assumption of naive expectations. This assumption was already applied in Ando and Modigliani, p. 57, ftn. 5 in an open model. In a closed version it was applied by Gale (1974) to a 3-period-lived consumer having an undiscounted Cobb-Douglas utility function. Kehoe and Levine (1985) also examined naive and rational expectations in a 2-cohort, multigood model. Unfortunately, these promising attempts were not followed. In contrast, the practitioners know quite well the naive expectations. For example, calculating the monthly mortgage payments, the bankers also assume that the recent interest rate remains unchanged, just to recalculate everything in the next year (Simonovits, 1992).

Dynamics

Obviously, now each consumer has to revise his consumption plan at each time-period. Consider the consumer born in $t - i$ when he is of age $i - 1$, that is, at the end of time-period $t - 1$.

We shall specify notations (C.15)–(C.16) to naive expectations (C.7). Dropping the index of calendar time:

$$(C.40) \qquad W_i(r) = \sum_{j=0}^{D-i} w_{i+j} r^{-j},$$

$$(C.41) \qquad V_i(r) = \sum_{j=0}^{D-i} \Phi^j r^{-\mu j}.$$

Substituting them into (C.17):

$$(C.42) \qquad c_i(a_{i-1}, r) = \frac{r a_{i-1} + W_i(r)}{V_i(r)}.$$

Remark. The fact that every consumer optimizes each period makes room for the reconsideration of other parameters as well. For example, the consumer of age i may revise his expectations on his forthcoming earnings and improve his decision rules. This problem, however, is beyond the scope of this book (see for example, Deaton, 1992).

We present now the simplest 3-cohort example:

Example C.6. Gale's (1974) example, naive expectations, undiscounted Cobb–Douglas utility. $D = 2$, $w_1 = 1$, $\mu = 0$, $\beta = 1$. By (C.42), $c_{0,t} = 1/(3r_t)$, $c_{1,t} = (r_t a_{0,t-1} + 1)/2$, $c_{2,t} = r_t a_{1,t-1}$. According to (C.19), $3(a_{0,t-1} + 2a_{1,t-1})r_t^2 - 3r_t + 2 = 0$. We solve the quadratic equation for r_t and choose the golden rule root. $r_t = 1$, that is, $c_{i,t} \equiv 1/3$, $a_{0,t-1} = -1/3$, $a_{1,t-1} = 1/3$.

First we assume that $A_{-1} = a_{0,-1} + a_{1,-1} = 0$, that is, $a_{0,t} + a_{1,t} = A_t = R_{-1,t} A_{-1} = 0$. Then our 2-dimensional difference equation reduces to one dimension: $a_{1,t} = [3 + \sqrt{9 - 24a_{1,t-1}}]/12$, $t = 0, 1, \ldots$. A simple computation shows that the system oscillates with increasing amplitude.

It is noteworthy that different initial states generate stable and unstable paths, (probably for $A_{-1} < 0$ and $A_{-1} > 0$), respectively.

Stability

What can be said on stability in general? We can determine the Jacobian matrix $K_F = (k_{ij,F})$ of the map f but for the time being, we cannot calculate or estimate in general its *spectral radius* (the maximum of the moduli of the characteristic roots), which determines stability.

We shall introduce the following notations (the numbers refer to the source equations).

$$(C.40a) \quad W_{i,F} = \sum_{j=0}^{D-i} w_{i+j} r_F^{-j},$$

$$(C.40b) \quad W'_{i,F} = -\sum_{j=0}^{D-i} j w_{i+j} r_F^{-j-1},$$

$$(C.41a) \quad V_{i,F} = \sum_{j=0}^{D-i} \Phi^j r_F^{-\mu j},$$

$$(C.41b) \quad V'_{i,F} = -\mu \sum_{j=0}^{D-i} j \Phi^j r_F^{-\mu j-1},$$

$$(C.17a) \quad c_{i,F} = \Phi^i r_F^{(1-\mu)i} \frac{W_{0,F}}{V_{0,F}},$$

$$(C.2a) \quad a_{i,F} = r_F a_{i-1,F} + w_i - c_{i,F}, \quad a_{-1,F} = 0,$$

$$(C.42b) \quad c'_{i,F} = \frac{(a_{i-1,F} + W'_{i,F})V_{i,F} - (r_F a_{i-1,F} + W_{i,F})V'_{i,F}}{V_{i,F}^2},$$

$$(C.18b) \quad S'_F = -\sum_{i=0}^{D} c'_{i,F},$$

$$(C.43) \quad k_{ij,F} = r_F \delta_{i-1,j} \left(1 - \frac{1}{V_{j+1,F}}\right) + r_F \frac{a_{i-1,F} - c_{i,F}}{V_{j+1,F} S'_F}$$

where δ_{ij} is the Kronecker symbol: it is equal to 1 if $i = j$ and 0 otherwise. Note that (C.42b) is a partial derivative of (C.42) but not the total derivative of (C.20a).

Lemma C.2. *Naive expectations. The Jacobi matrix of $f(a)$ around the steady state a_F is given by $K_F = (k_{ij,F})$.*

Proof. We linearize system (C.2) around its fixed point a_F where $r(a)$ is determined by (C.42) and (C.18):

$$k_{ij} = \frac{\partial r}{\partial a_{i-1}} + r \delta_{i-1,j} - \frac{\partial c_i}{\partial a_j}.$$

The first partial derivative can be determined from the implicit function (C.19): $S(a,r) = 1 - \sum_{i=0}^{D} c_j(a_{j-1}, r)$ yields

$$\frac{\partial r}{\partial a_j} = -\frac{\partial S/\partial a_j}{\partial S/\partial r} = -\frac{r}{V_{j+1} S'}.$$

The second partial derivative is

$$\frac{\partial c_i}{\partial a_j} = \delta_{i-1,j} \frac{r}{V_{j+1}} + c'_i \frac{\partial r}{\partial a_j}.$$

Substitutions give (C.43). ∎

At this point we make a helpful observation on the *singularity condition* $S'_G = 0$. If $\beta = 1$, then $V_{i,G}$, $V'_{i,G}/\mu$, $c_{i,G}$, $a_{i,G}$ are independent of μ. Thus (C.18b) reduces to $S'_G = \pi - \pi_\mu \mu$ where π and π_μ are two parameters which are independent of μ. Thus $S'_G = 0$ is equivalent to $\mu_o = \pi/\pi_\mu$. Obviously, if $\mu_o < 0$ or $\mu_o > 1$, then there is no singularity problem, at least for $\beta = 1$ and $r_G = 1$. If $0 \leq \mu_o \leq 1$, then the model may have no solution for $\mu = \mu_o$. In addition, around this value, instability might be expected. It is difficult to say anything general, but numerical calculations show that for $M = 55$, $\mu_o = 0.742$.

To examine the stability of our system, we use a well-known stability criterion (Theorem 1.4).

Remarks. 1. For $\rho(K_F) = 1$, further examinations are needed to determine if the system is stable, unstable or Lyapunov stable.

2. Stability is only a qualitative property of dynamics. It would be interesting to characterize the dampening factor. Apart from multiplicity and complex roots, the deviations $a_{i,t} - a_{i,F}$ asymptotically form a decreasing geometric series: $a_{i,t} - a_{i,F} \approx (a_{i,t-1} - a_{i,F})\rho(K_F)$, thus $1/\rho(K_F)$ gives the *dampening factor* in the asset dynamics. It would be desirable to apply this criterion the same way as we applied Lemma C.1 to Theorem C.7. Presently we have only fragmentary results. For example, we can prove a remarkable property.

Lemma C.3. *Naive expectations.* a) *If $0 < -A_G < S'_G$, then the maximal column sum of matrix K_G is equal to $K_{D-1,G} = 1 + A_G/S'_G$.*

b) *The column sums of matrix K_B are all equal to r_B.*

Proof. First we shall consider any feasible steady state. Note that the denominators of entries in column j are the same. Take the numerators. By (C.2a) and (C.42b), $c'_{D,F} = a_{D-1,F}$, hence via (C.18b) $S'_F = -\sum_{i=0}^{D-1} c'_{j,F} - a_{D-1,F}$. Similarly, $A_F = a_{D-1,F} + \sum_{i=0}^{D-1} a_{j-1,F}$, thus $K_{j,F} = r_F(1 + A_F/(V_{j+1,F}S'_F))$. a) For the golden rule steady state, $\{V_{j+1,G}\}$ is a decreasing sequence, thus under our condition, $0 < K_{j,G} \leq K_{D-1,G} = 1 + A_G/S'_G$. b) For a balanced steady state, $A_B = 0$ implies $K_{j,B} = r_B$. ∎

Lemma C.3 can be used at the discussion of several interesting issues.

Example C.7. Naive expectations, $D + 1$ cohorts, no discount, autarky. Let $L = 0$, $M = D$, $\beta = 1$, $w_i \equiv 1/(D+1) \equiv c_{i,G}$. Then $a_{i,G} = 0$, $c'_{i,G} = (1-\mu)(i-D)/[2(D+1)]$. Here the singularity value is given as $\mu_o = 1$. Obviously, $k_{ij,G} > 0$, and Lemma C.3 implies $K_{j,G} = 1$, $j = 0, \ldots, D-1$. The Perron-theorem (Theorem A.8b) now implies that $\rho(K_G) = 1$, which is just that case where Theorem 3.3 does not tell anything on stability. Probably the symmetric steady state is stable from the left. Nevertheless, Example C.7 is an appropriate starting point to

Theorem C.8. *Naive expectations. Let $\mu < 1$, let β and w_i be sufficiently close to 1 and $1/(D+1)$, respectively, that $K_F > 0$.*
a) If $r_B > 1$, then the debtor golden rule steady state is stable.
b) If $r_B < 1$, then the balanced steady state is stable.

Remarks.   1. It would be interesting to know the range of applicability of Theorem C.8. Example 2 in Molnár and Simonovits (1996) shows maximum validity for $D = 1$. Computer simulations reveal that for $L = 0$, $M = D$, $0.98 \leq \beta \leq 1$ and $w_i = w_0 \Omega^i$, $0.99 \leq \Omega \leq 1.01$, $K_F > 0$ holds. The 3-cohort Cobb–Douglas Example C.6, which is due to Gale (1974), is a case in point when simple analysis demonstrates that K_F has both positive and negative entries, thus for $D > 2$, stronger methods are needed.
2. In the light of Theorem C.8, at the symmetric steady states in Example C.7 one-sided stability prevails.

Proof.   a) According to Theorem C.5a, now $\mu_1 = 1$, thus $A_G < 0$ holds. By continuity, $S'_G > 0$ also holds, implying $0 < K_{j,G} < 1$ for $j = 0, \ldots, D - 1$. b) $K_{j,B} = r_B$ for $j = 0, \ldots, D - 1$. In both cases, the Perron-theorem implies $\rho(K_F) < 1$, hence Theorem 1.5 applies. ∎

What happens if we renounce the assumptions of near autarky, moreover, we allow zero earnings? For the time being, a more general and realistic analysis requires a computer. For selected parameter vectors, computations showed that $\rho(K_B) = r_B$ (with a left hand positive characteristic vector $\mathbf{1^T}$), although each K_B has both positive and negative entries.

According to Theorem 1.6, the dampening factor is equal to the reciprocal of the balanced interest factor. However, $\rho(K_G)$ cannot be estimated by $K_{D-1,G}$ anymore. In fact, in several cases the dominant characteristic roots are complex conjugates. For one case, two steady states are stable (see Example C.8 below).

Conjecture C.2. *Naive expectations. For realistic models, typically at least one steady state is stable.*

Viability

Our stability results are very useful but tell us very little on the size of the neighborhood in which the paths are really convergent. Nor do they inform us on the transient behavior, that is, on the viability of the system. Therefore we made some computer runs which display always smooth but sometimes divergent paths.

Unlike rational expectations, naive expectations reproduce steady states without much trouble. What happens if the system is disturbed? Assume that the initial interest factors are not at the steady state. For simplicity, let us assume that they are given as in the aggregator map, that is,

$$r_t = r, \quad c_{i,t}(r) = \Phi^{t-i} r^{(1-\mu)(t-i)} \frac{W_i(r)}{V_i(r)} \quad (1 \le i \le -t < D).$$

For slightly distorted initial values, naive expectations behave quite smoothly, at least between the two steady states and slightly below the lower steady state.

Finally, we are trying to answer what happens to the stability in the window of Theorem C.5c.

Example C.8. Naive expectations, dual stability. Continuation of Example C.4. Note that there is a second balanced steady state, around $r = 0.3$. Figure C.3 shows three paths again. The paths starting from 0.92 and 0.94 converge to r_B while the paths starting from 0.995 converges to 1. Note that the corresponding aggregator map (Figure C.4) has two branches. The lower branch determines the balanced steady state, while the upper branch cuts out the golden rule one. This is the reason that two neighboring steady states are stable.

Remark. It is an open question whether cycles can be produced by jumping from either branch to the other.

Having presented our analysis, we have only few concluding remarks to make. (i) The practice of lumping many cohorts together in generations is misleading: while aggregate models may display stability and cycles for certain parameter values, apparently independent of aggregation; disaggregated models often display instability and inviability. (ii) Rational expectations may be very elegant, but in the multicohort framework the generated dynamics is so unstable that even the steady states can hardly be simulated. (iii) Naive expectations may appear very stupid, but, contrary to rational expectations, they function quite well in OLC models. Our research is just a start in new directions. Much further work needs be done to clarify the foregoing issues.

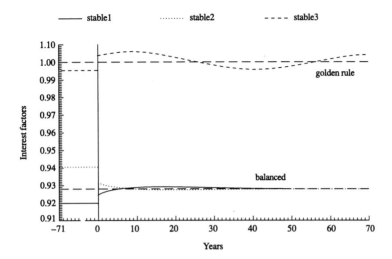

Figure C.3
Dual stability for naive expectations

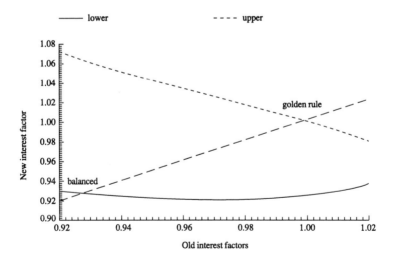

Figure C.4
Aggregator maps for naive expectations

SOLUTIONS

Chapter 1

Problem 1.1. $x_{1,t} = g(x_{1,t-1}, x_{2,t-1}) = -x_{2,t-1}$ and $x_{2,t} = x_{1,t-1}$.

Problem 1.2. Because $y_0 \neq 0$.

Problem 1.3. a) No, because the deviations do not converge to zero. b) Yes, because the deviations remain bounded.

Problem 1.4. a) $P = 4$ and b) $y_t = -y_{t-1}$, $y_0 = -1$: $P = 2$.

Problem 1.5. From (1.13)

$$(I - M)^{-1} = \delta \begin{pmatrix} 1 & \alpha \\ \beta & 1 \end{pmatrix}, \qquad x^\circ = \delta \begin{pmatrix} 1 + \alpha \\ 1 + \beta \end{pmatrix}$$

where $\delta = 1/(1 - \alpha\beta)$.

Problem 1.6. A characteristic vector s and the corresponding characteristic value λ are determined by $\lambda s = Ms$, that is, coordinatewise $\lambda_j s_{1,j} = \alpha s_{2,j}$ and $\lambda_j s_{2,j} = \beta s_{1,j}$. After multiplication: $\lambda_j^2 s_{1,j} s_{2,j} = \alpha\beta s_{1,j} s_{2,j}$. No characteristic vector may have coordinate zero, because then the other coordinate would also be zero. Therefore $\lambda_j^2 = \alpha\beta$. Thus we may normalize as $s_{1,j} = 1$, that is, $\lambda_j s_{2,j} = \beta$, that is, $s_{2,1} = \sqrt{\alpha/\beta}$ and $s_{2,2} = -\sqrt{\alpha/\beta}$. From (1.20) $x_0 = \xi_1 s_1 + \xi_2 s_2$, whence the given initial state $x_0 = (x_{1,0}, x_{2,0})$ yields ξ_j. (1.21) provides the solution.

Problem 1.7. First we solve the scalar equation $x_{1,t}$ $= m_{11}x_{1,t-1} + w_1$. We substitute the solution obtained into the scalar equation $x_{2,t} = m_{22}x_{1,t-1} + m_{21}x_{1,t-1} + w_2$, and so on. The only complication arises from the exponential terms on the R.H.S., which can be tackled, see (1.36).

Problem 1.8. Column j of matrix M shows the image of the unit vector e_j, thus magnification and rotation are combined. The characteristic equation is $P(\lambda) = \lambda^2 - 2\rho(\cos\varphi)\lambda + \rho^2 = 0$ [(1.27)–(1.28)], yielding the foregoing characteristic roots. The geometric meaning also suggests this result.

Problem 1.9. According to Problem 1.6, M_1 has two characteristic values: 1 and –1. According to Problem 1.7, both characteristic values of M_2 are equal to 1, however, there exists only a single independent characteristic vector: $s = (0,1)$.

Problem 1.10. $x_t = \lambda x_{t-1} + w$ where x_t, λ and w are scalars. Theorem 1.1: if $\lambda \neq 1$, then $x^\circ = w/(1 - \lambda)$. Theorem 1.2: trivial. Theorem 1.3: empty. Theorem 1.4: stability $-1 < \lambda < 1$. Theorem 1.5: $0 \leq \lambda \leq 1$. Theorem 1.6: $\Phi = 1/|\lambda|$. Theorem 1.7: $x_t = x_{t-1}$ is meaningless. Theorem 1.8 reduces to Theorem 1.4.

Problem 1.11. a) $\lambda_1 = \lambda_2 \geq 0$. b) at least one sign-change, at least two sign-changes.

Problem 1.12. a)

$$B^{-1} = \begin{pmatrix} 0 & 1/\beta \\ 1/\alpha & 0 \end{pmatrix}.$$

b) $b_{11} = 0$. c) Example 1.15. d) Similarly to Problem 1.6, the characteristic equation of matrix $I - B\langle k \rangle$ is $P(\lambda) = (\lambda - 1)^2 - \alpha\beta k_1 k_2 = 0$, whence $\lambda = 1 \pm \sqrt{\alpha\beta k_1 k_2}$, $\lambda_1 \geq 1$.

Chapter 2

Problem 2.1. Programming.

Problem 2.2. Insert the stationary values into (2.18)–(2.19). Equations $y^\circ = Y^\circ 1 + c$ and $Y^\circ = A\langle y^\circ \rangle$ yield the traditional equation $y^\circ = Ay^\circ + c$, whence $y^\circ = (I - A)^{-1}c$, $Y^\circ = A\langle y^\circ \rangle$. Considering the stationary conditions (2.20)–(2.21): system $y^\circ + \langle d \rangle z^\circ = y^*$, $Y^\circ + D \times Z^\circ = Y^*$ has a positive solution in (z°, Z°), assuming that $y^\circ < y^*$ and $Y^\circ < Y^*$ hold.

Problem 2.3. For $|\lambda| = 1$, expression

$$|\pi(\lambda)|^2 = \frac{\beta^2 + \alpha^2 - 2\alpha\beta\Re\lambda}{1 + \varepsilon^2 - 2\varepsilon\Re\lambda}$$

is defined for all $-1 \le \Re\lambda \le 1$. The hyperbolic function $|\pi(\lambda)|^2$ reaches its maximum at one end point. Since $\pi(1) = (\beta - \alpha)/(1 - \varepsilon)$ and $\pi(-1) = (\beta + \alpha)/(1 + \varepsilon)$, under our assumptions, $0 \le \pi(1) < \pi(-1)$ hold.

Problem 2.4. (See Lovell, 1962 and Martos, 1990.) For uniform norms and reactions, $b = (1 + \gamma\varepsilon)\mathbf{1} = \beta\mathbf{1}$ holds, that is, matrix (2.43) is a rational function of N, that is, λ is a likewise function of ν. $\lambda = 1 - (1 - \nu)\varepsilon - (1 - \nu)\beta(1 - \nu\beta)^{-1}\nu\varepsilon$.

Problem 2.5. Let now $\beta_i = b_i + k_i$, $\alpha_i = b_i$ and $\varepsilon_i = 1 - k_i$, $\pi_i = p_i$, $i = 1, \ldots, n$. On the basis of Problem 2.3 one can prove again that, due to $\pi_i(-1) \le 0$, $\rho[N\langle -p(-1)\rangle] \le 1$ is sufficient but generally not necessary. If, however, N is 2-cyclic, then $-\rho(N)$ is also a dominant characteristic value, that is, $\rho[N\langle -p(-1)\rangle] \le 1$ is also necessary.

Chapter 3

Problem 3.1. a) $f(x) > x$ holds, though $0 < f'(x) = 1 - (1 + e^x)^{-2}e^x < 1$. b) Interval $(-\infty, \infty)$ is not compact.

Problem 3.2. a) The map is indeed not a contraction, since for $\beta = 1$ and $x_0 = 0.01$, $x_1 = 50.005$, while for $y_0 = 1$, $y_1 = 1$.

b) Comparing the arithmetic and geometric means yields $x_t \ge \sqrt{\beta}$ $(t \ge 1)$. Then $0 < f'(x) = 1/2 - \beta/(2x^2) < 1$, that is, contraction. Note that $f'(\sqrt{\beta}) = 0$, that is, the convergence is very fast.

Problem 3.3.* Draw the diagram known from the cobweb -cycle. After some trials one can discover the following case by case approach.

a) Figure 3.1: If $1 \le a < 2$, then the fixed point $f(x^\circ) = x^\circ$ lies in $(0; 1/2)$. Due to symmetry, for $x^* = 1 - x^\circ$, $f(x^*) = x^\circ$. If $x^* < x_0 < 1$, then $x_1 < x^\circ$; if $1/2 < x_0 < x^*$, then $x^\circ < x_1 < 1/2$. We have demonstrated that the system can stay to the right from $1/2$ for at most one time-period. If $x^\circ < x_t < 1/2$, then $x^\circ < x_{t+1} < x_t$;

if $0 < x_t < x^\circ$, then $x_t < x_{t+1} < 1/2$. Because of monotonicity, the limit exists, which is naturally a fixed point, that is, is equal to x°.

b) Figure 3.2: If $2 < a < 3$, then there exists such a number $0 < x_1^* < 1/2$, for which $f(x_1^*) = 1/2$. Because of symmetry, $x_2^* = 1 - x_1^*$, $f(x_2^*) = 1/2$. If $x_2^* < x_0 < 1$, then $x_1 < x_1^*$. If $1/2 < x_0 < x_2^*$, then $x_1^* < x_1 < 1/2$. If $x_t < x_1^*$, then $x_t < x_{t+1} < x_1^*$. If $x_1^* < x_t < 1/2$, then $x_t < x_{t+1} < x_2^*$; that is, sooner or later the system arrives to the open interval (x_1^*, x_2^*) and stays there. Then the contraction map theorem is applicable, because

$$x_{1,2}^* = \frac{1 \pm \sqrt{1 - a/2}}{2} \Rightarrow 0 \le f'(x_1^*) = a(1 - 2x_1^*) = \sqrt{1 - 2/a} < 1.$$

c) For $a = 2$, there is a borderline case between a) and b).

Problem 3.4.* If matrix A (or B) is invertible, then the proof of the Lemma is elementary: $ABx = \lambda x$, $y = A^{-1}x \Rightarrow BAy = \lambda y$. Hence via a limit-argument the Lemma applies for any matrix A. Thus $\rho(AB \cdots C) = \rho(B \cdots CA)$, that is, matrix $\mathbf{D}f^P(x_2) = \mathbf{D}f(x_1)\mathbf{D}f(x_P) \cdots \mathbf{D}f(x_2)$ is stable.

Problem 3.5. a) $x^\circ = 2 - 2x^\circ \Rightarrow x^\circ = 2/3$. b) $x_2 = 2x_1$ and $x_1 = 2 - 2x_2 = 2 - 4x_1 \Rightarrow x_1 = 2/5 \Rightarrow x_2 = 4/5$. c) Out of three, either two points are less than $1/2$ or two points are greater than $1/2$. Then continue as in b). d) Unstable, because $|f'(\cdot)| \ge 2$, or $4, 8 > 1$.

Problem 3.6. $\sin 2\varphi = 2 \sin \varphi \cos \varphi$ implies $\varphi_t = 2\varphi_{t-1}$, and so on.

Problems 3.7–3.9. Programming.

Chapter 4

Problems 4.1–4.3. Programming.

Problem 4.4. In fact, we used (4.28′) rather than the time-invariance of A_t at (4.29).

Problem 4.5.* Introducing the general control-theoretic framework with bounds, we have the state space equation $x_t = x_{t-1} + Bu_t$, and the planned control equation $u_t^p = -\langle k \rangle x_t$ and the actual control equation

$$u_{i,t} = \begin{cases} u_i^l & \text{if } u_{i,t}^p \le u_i^l; \\ u_{i,t}^p & \text{if } u_i^l < u_{i,t}^p < u_i^u; \\ u_i^u & \text{if } u_{i,t}^p \ge u_i^u; \end{cases} \quad i = 1, \ldots, n.$$

where u_i^l and u_i^u are lower and upper control bounds, respectively.

Similarly to Section 2.2, we can generalize the analysis not only to the extended stock signalling model but to any Metzlerian control model.

Chapter 5

Problem 5.1. Because of $f(t, x) = \lambda x$, with notation $h = t/k$, the

$$x_k(t_{i,k}) = x_k(t_{i-1,k}) + \lambda x_k(t_{i-1,k})h = (1 + \lambda h)x_k(t_{i-1,k}).$$

According to the formula of the geometric series, the solution is $x_k(t)$ $= x_k(0)(1 + \lambda t/k)^k$, therefore for $k \to \infty$, $x_k(t)$ converges to $e^{\lambda t}$.

Problem 5.2. Due to $f[\tau, x(\tau)] = \lambda x(\tau)$, the kth integral equation is

$$x_{k+1}(t) = x(0) + \lambda \int_0^t x_k(\tau)\, d\tau.$$

For approximation 0, $x_0(t) \equiv 1$, the formula holds. Mathematical induction: Plug in approximation k and integrate by terms: the integral of term j is equal to $\lambda \lambda^j t^{j+1}/[(j+1)j!]$, which is just term $j+1$, $j = 0, 1, \ldots, k$, and term 0 remains 1.

Problem 5.3. For function $f(x) = x^{2/3}$, the difference-quotient $[f(x) - f(0)]/(x - 0) = x^{-1/3}$ is unbounded around $x \approx 0$.

Problem 5.4. a) $dx/dt = \lambda x \Rightarrow x^{-1}dx = \lambda dt \Rightarrow \log x = \lambda t + c \Rightarrow x(t) = e^{\lambda t + c} \Rightarrow x(0) = e^c \Rightarrow x(t) = x_0 e^{\lambda t}$.

b) $dx/dt = x^{2/3} \Rightarrow x^{-2/3}dx = dt \Rightarrow 3x^{1/3} = t \Rightarrow x(t) = (t/3)^3$.

c) $dx/dt = x^2 \Rightarrow x^{-2}dx = dt \Rightarrow -x^{-1} = t + c \Rightarrow x(t) = -1/(t + c) \Rightarrow x(0) = -1/c$ and so on.

Problem 5.5. Using (5.17*)–(5.18*) and Theorem 5.7, we have $\lambda = i$ and $s = (1, -i)/2$. Then $\Re s = (1, 0)/2$ and $\Im s = -(0, 1)/2$.

Problem 5.6. We shall apply Theorem 5.8. According to the relation among the roots and coefficients, $\lambda_1 + \lambda_2 = -\alpha$ and $\lambda_1\lambda_2 = \beta$. a) If a system is stable, then the real parts of both roots are negative, that is, $\alpha > 0$ and $\beta > 0$ (either real or complex). b) Assume $\alpha > 0$ and $\beta > 0$. If both roots are real, then neither one root nor two roots can be positive (because $\beta > 0$ and $\alpha > 0$, respectively). If both roots are complex conjugates, then their real parts are equal to $-\alpha/2$, that is, negative, that is, the system is stable.

Chapter 6

Problem 6.1. $Y_t = C_t + I_t$, $Y_t = Y_{t-1} + AI_{t-1}$, $C_t = (1-s)Y_t \Rightarrow$ $Y_t = (1 + As)Y_{t-1}$.

Problem 6.2.

Problem 6.3. (i) Varian (1992, Chapter 8.) (ii) Indirect: let $p^* \neq \pi p^\circ$ ($\pi > 0$) another equilibrium price vector. Since $z(p^*) \leq 0$, $z(p^*) \neq 0$ and $p^\circ > 0$, $p^{\circ T} z(p^*) < 0$, contradiction.

Problem 6.4. We have a single independent variable and a unique equilibrium, thus the solution of $\dot{p}_1 = f(p_1)$ converges monotonously to the equilibrium.

Problem 6.5. Let us shift the center of the ellipsoid introduced in (6.34) to the equilibrium price vector and consider the resulting Lyapunov function

$$V(p) = \sum_{i=1}^{n} \frac{[p_i - p_i(0)]^2}{d_i}.$$

Chapter 7

Problem 7.1. $v_1(x_1) = \min[x_1^2 + u_1^2 + x_2^2 | u_1] = \min[x_1^2 + u_1^2 + (x_1 + u_1)^2 | u_1]$. Taking the derivative of [ ] with respect to u_1, $u_1 + x_1 + u_1 = 0$, that is, $u_1 = -x_1/2$ obtains. Inserting back: $v_1(x_1) = 3x_1^2/2$. $v_0(x_0) = \min[x_0^2 + u_0^2 + v_1(x_1) | u_0] = \min[x_0^2 + u_0^2 + (3/2)(x_0 + u_0)^2 | u_0]$. Taking the derivative of [ ] with respect to u_0, $2u_0 + 3(x_0 + u_0) = 0$ results in $u_0 = -(3/5)x_0$.

Problem 7.2. $x_{t+1} = x_t + u_t$, $x_0 = -1$, $x_T = 0$. $A_t = B_t = G_t = 1$, $F_t = 0$, $t = 0, 1, \ldots, T-1$ and $F_T = 0$. Computation demonstrates that $S_t = S_{t+1}/(S_{t+1} + 1)$ and $K_t = S_t$. By induction we obtain $S_t = 1/(T-t)$, $x_t = (T-t)x_0/T$, $u_t = 1/T$.

Problem 7.3. We shall reduce the problem to Example 7.3. Introduce $x = \log \xi$ and $y = \log \eta$. Substituting into the functional equation: $V(e^{x+y}) = V(e^x e^y) = V(e^x) + V(e^y)$ and introducing $v(x) = V(\log x)$, yields $v(x + y) = v(x) + v(y)$ with $v(1) = V(e) = 1$.

Problem 7.4. Note that $\mathbf{E}(x_1 + u_1 + w_1)^2 = (x_1 + u_1)^2 + \mathbf{E}w_1^2$.
$v_1(x_1) = \min[x_1^2 + u_1^2 + \mathbf{E}x_2^2|\ u_1] = \min[x_1^2 + u_1^2 + (x_1 + u_1)^2 + \mathbf{E}w_1^2|\ u_1]$.
Taking the derivative of $[\ ]$ with respect to u_1, $u_1 + x_1 + u_1 = 0$
yields $u_1 = -x_1/2$. Plugging back: $v_1(x_1) = 3x_1^2/2 + \mathbf{E}w_1^2$. $v_0(x_0) = \min[x_0^2 + u_0^2 + v_1(x_1)|\ u_0] = \min[x_0^2 + u_0^2 + (3/2)\mathbf{E}(x_0 + u_0 + w_0)^2 + \mathbf{E}w_1^2|\ u_0]$. Taking the derivative of $[\ ]$ with respect to u_0, $2u_0 + 3(x_0 + u_0) = 0$ results in $u_0 = -(3/5)x_0$.

Chapter 8

Problem 8.1. See Section 10.2 or C.2.

Problem 8.2. Assume that the government introduces a mandatory pension system where annuity b is determined from the equality of expected future values of the capital and of pensions: $r^2k_0 = rb + qb$, hence

$$b = \frac{r^2k_0}{r+q} = c\frac{r+1}{r+q} > c.$$

Problem 8.3. $\sum_{t=0}^{\infty}\beta^t c_t \to$ max, subject to $0 \le c_t \le k_t - \beta k_{t+1}$, $t = 0, 1, \ldots$, k_0 given. $k_t = \beta k_{t-1} = \cdots = \beta^t k_0$ does not satisfy the transversality condition.

Problem 8.4. $\sum_{t=0}^{\infty}\beta^t(k_t - \beta k_{t+1}) \to$ max, subject to $0 \le k_{t+1} \le k_t/\beta$, $t = 0, 1, \ldots$, k_0 given. $k_t = \beta k_{t-1} = \cdots = \beta^t k_0$ does not satisfy the transversality condition.

Problem 8.5. a) Take the reciprocal of (8.8) and multiply by Ψ and divide by k_t: $1 - k_{t+1}/k_t^{\alpha} = \Psi k_t/k_{t-1}^{\alpha} - \Psi$. It is almost plausible that z_t is to be chosen. b) Trial and error.

Problem 8.6. Substitution.

Problem 8.7. $\sum_{i=0}^{t}\alpha^i \log \Psi$.

Problem 8.8. If the initial value $k_0 = [1 + \cos(3\pi/8)]/2$ is chosen, then $k_1 = [1 + \cos(\pi/4)]/2$, $k_2 = 1/2$, $k_3 = 1$ and $k_4 = k_5 = \cdots = 0$. Inserting the obtained finite path into (8.17), the following cubic equation is obtained:

$$\beta^{*3} - \frac{\beta^{*2}}{2} + \frac{\beta^*}{2}\cos\frac{\pi}{4} - \sin\frac{\pi}{16}\sin\frac{5\pi}{16} = 0,$$

its positive root is approximately equal to 0.475.

Problems 8.9–8.10. See Levhari and Mirman (1980).

Chapter 9

Problem 9.1. $H(x, u, p) = f(x, u) + p^{\mathbf{T}} g(x, u)$. Take the total time derivative of H: $dH/dt = H_x \dot{x} + H_u \dot{u} + H_p \dot{p}$. Taking into account $\dot{x} = H_p^{\mathbf{T}}$; $\dot{p} = -H_x^{\mathbf{T}}$ and $H_u = 0$, yields $dH/dt = 0$.

Problem 9.2. a) Triangle inequality: $\overline{AB} + \overline{BC} \geq \overline{AC}$. This plausible fact is proved by the more sophisticated lemma: *in a triangle, greater angle corresponds to longer side.* Assume that $\overline{AC}$ is the longest side of the triangle. Measure interval $\overline{BC}$ to the point B of $\overline{AB}$. In the resulting triangle $AC'C$ at vertex C' angle γ, at vertex C angle $\gamma + \gamma'$ can be found, that is, according to the Lemma, $\overline{AC} \leq \overline{AC'} = \overline{AB} + \overline{BC}$.

b) $f(t, x, \dot{x}) = \sqrt{1 + \dot{x}^2} \geq 1$. Note, how difficult is to apply the Euler–Lagrange differential equation: Using the second case of deficient reward functions (discussed in Section 9.3), $f_{\dot{x}}(t, \dot{x}) = c$, that is, $\dot{x}/\sqrt{1 + \dot{x}^2} = c$, hence $\dot{x} = k$, thus $x \equiv 0$.

Problem 9.3. Reflect point I to the line of the mirror: I' and denote by T the tangent point. $\overline{OT} + \overline{TI} = \overline{OT} + \overline{TI'}$. By Problem 9.2, then the total distance is minimal if T lies on the line connecting O to I'.

Problem 9.4. Let the border water-air be the axis t, let the eye and the object be point $(0, a)$ of axis x and point (b, d) of the positive half plane, respectively. Let $(0, x)$ be the refraction point, u and v the reciprocal of the speed of the light in the air and the water, respectively. Then the time-minimum-problem is as follows: $u\sqrt{a^2 + x^2} + v\sqrt{d^2 + (b - x)^2}$. Taking the derivative of the last expression with respect to x, we obtain $u[a^2 + x^2]^{-1/2}x - v[d^2 - (b - x)^2]^{-1/2}(b - x) = 0$, which provides the foregoing law.

Problem 9.6.* a) In abscissa t the elementary mass $\sqrt{1 + \dot{x}(t)^2}dt$ has potential energy $x(t)\sqrt{1 + \dot{x}(t)^2}dt$. b) Inserting to the deficient form (i): $(x - p)(1 + \dot{x}^2)^{1/2} - \dot{x}^2(x - p)(1 + \dot{x}^2)^{-1/2} = c$. After rearrangement: $\dot{x} = \sqrt{\dot{x}(t)^2 - C^2}/C$, hence $x(t) = A\cosh(\delta t) + \gamma$. Length: $2A\sinh(\delta a) = \kappa$, determines the value of δ. End point: $D = A(e^{-\delta a} + e^{\delta a})/2 + \gamma$ determines the value of γ. c) The deepest point of the curve is at abscissa $t = 0$: $x(0) = A + \gamma \geq 0$.

Chapter 10

Problem 10.1. $\delta = 0$, thus from (10.11) $\dot{c} = 0$, hence $c(0) = (k_0 + wT)/T = 1.25$, a uniform running through the capital.

Problem 10.2. By $f'(k^\circ) = \beta$, $A\alpha k^{\alpha-1} = \beta$, that is, $k^\circ = [A\alpha/\beta]^{1/(1-\alpha)}$ and $c^\circ = f(k^\circ) = A(k^\circ)^\alpha$. $r^\circ = \beta$. The characteristic equation is $\lambda^2 - \beta\lambda - q^\circ = 0$ where $q^\circ = -f''(k^\circ)/a^\circ = (1-\alpha)\beta c^\circ/(\varepsilon k^\circ) \Rightarrow \lambda_1 = [\beta - \sqrt{\beta^2 + 4q^\circ}]/2$.

Problem 10.3. $k^\circ = 142.8$; $c^\circ = 44.3$; $r^\circ = 0.03$; $\lambda_1 = -0.034$.

Problem 10.4. Slow convergence, see Figures 10.1–10.2. It is remarkable that even if the system starts from the equilibrium, the discrete steps remove the system from the equilibrium (see Example 1.10.)

Appendix A

Problem A.1. Return to the beginning of Section 1.3. Let $m_{11}, m_{22} \geq 0$ in (1.28). Because M is irreducible, $m_{12}, m_{21} > 0$.
a) To have a positive characteristic value, it is necessary that both characteristic values be real. We have to check if $\omega^2 \geq 4\vartheta$ [(1.29)] holds. Yes, because substituting (1.28) into (1.29), yields $(m_{11} - m_{22})^2 + 4m_{12}m_{21} \geq 0$. Since the sum of the two characteristic values $(-\omega)$ is nonnegative, and the discriminant is positive; there exists a positive characteristic value, which is (a) dominant characteristic value.

b) Rearranging the characteristic equation: $(\lambda - m_{11})x_1 = m_{12}x_2$, $(\lambda - m_{22})x_2 = m_{21}x_2$. Multiplying and simplifying with $x_1 x_2 \neq 0$: $(\lambda - m_{11})(\lambda - m_{22}) = m_{12}m_{21} > 0$. Assume $m_{11} \geq m_{22}$. Then $\lambda_2 \leq m_{22} \leq m_{11} \leq \lambda_1$. For a characteristic vector belonging to a dominant positive root, $x_1/x_2 = m_{12}/(\lambda_1 - m_{11}) > 0$. The other characteristic vector satisfies $y_1/y_2 = m_{12}/(\lambda_2 - m_{11}) < 0$.
c) Since the discriminant is positive, the two characteristic values are different.
d) $2\rho(M) = m_{11} + m_{22} + \sqrt{(m_{11} - m_{22})^2 + 4m_{12}m_{21}}$, that is, $\rho(M)$ is an increasing function of m_{12} and of m_{21}. Since the square root function is concave, in the interval $0 \leq m_{11} \leq m_{22}$, under the increase of m_{11} the discriminant grows slower than $|m_{11} - m_{22}|$.

e) With the help of the adjoint matrix, the inverse of $(I - M)$ is as follows:

$$(I - M)^{-1} = P(1)^{-1} \begin{pmatrix} 1 - m_{22} & m_{12} \\ m_{21} & 1 - m_{11} \end{pmatrix}$$

where $P(\lambda) = (\lambda - m_{11})(\lambda - m_{22}) - m_{12}m_{21}$ is the characteristic polynomial of M. If M is stable, then $-1 < \lambda_2 \leq \lambda_1 < 1$ and $P(1) > 0$. From b) $m_{ii} < \lambda_1 < 1$, thus $(I - M)^{-1} > 0$.

Problem A.2. Problem A.1, $P = 2$ if and only if $m_{11} = m_{22} = 0$. Then $\lambda_2 = -\lambda_1 = \sqrt{m_{12}m_{21}}$.

Problem A.3. If both column sums are identical, then this value is a characteristic root, say, $\lambda_1 = m_{11} + m_{12} = m_{12} + m_{22}$, with the corresponding left hand characteristic vector $(1, 1)$. The other characteristic value is $\lambda_2 = \det M / \lambda_1 = [m_{11}m_{22} - (\lambda_1 - m_{11})(\lambda_1 - m_{22})]/\lambda_1 = m_{11} + m_{22} - \lambda_1$. Hence the dominance condition is $0 \leq m_{11} + m_{22} \leq 2\lambda_1$. (If $M \geq 0$, then the assumption holds, since $m_{11}, m_{22} \leq \lambda_1$.)

Problem A.4. l_1: diamond with the following vertices: $(1, 0)$, $(0, 1)$, $(-1, 0)$, $(0, -1)$; l_2: unit disk with center 0, l_∞: square with the following vertices: $(1, 1)$, $(1, -1)$, $(-1, -1)$, $(-1, 1)$.

Problem A.5. Because of symmetry, only the case $\|M\|_\infty$ is considered. We shall apply the triangle inequality to $y_i = \sum_{j=1}^n m_{ij}x_j$: $|y_i| \leq \sum_{j=1}^n |m_{ij}| \, |x_j|$. Using the definition of $\|x\|_\infty$ and $\|M\|_\infty$, $|y_i| \leq \sum_{j=1}^n |m_{ij}| \|x\|_\infty \leq \|M\|_\infty \|x\|_\infty$.

It can be shown that our inequality is sharp. For example, let 1 be such a row index for which $\sum_{j=1}^n |m_{1j}| = \|M\|_\infty$ and let β_j be a complex number with unit modulus such that $|m_{1j}| = m_{1j}\beta_j$, $j = 1, \ldots, n$. Finally, with the choice $x_j = 1/\beta_j$, $|y_1| = \sum_{j=1}^n |m_{1j}|$ and so on.

Appendix B

Problem B.1. Plugging $c_{0,t} = c_{1,t+1}$ and $s_{i,t} = w_i - c_{i,t}$ into (B.3), yields the formulas.

Problem B.2. Standard microeconomics, see also (C.14)–(C.20).

Problem B.3. An increasing function has a positive derivative, the derivative of a strictly increasing convex function is increasing, Rolle-theorem implies the existence of such a point between two steady states where the derivative is equal to 1, that is, the derivative of the function at the lower steady state is between 0 and 1, while it is higher than 1 at the higher steady state.

Problem B.4. a) Note that in (B.4) r_{t+1} is replaced by r_t, that is, $s_{0,t} = s(r_t)$. b) However, in (B.3) r_{t+1} remains: $s_{1,t+1} = -r_{t+1}s(r_t)$. Again using the short-cut, $s_{0,t+1} = s(r_{t+1})$, hence at $t+1$, $S(r_t, r_{t+1}) = \nu s(r_{t+1}) - r_{t+1}s(r_t) = 0$. Local stability analysis yields the result.

Problem B.5. Apply the result of Problem B.2 to determine ε. Compare (B.8) and Problem B.4.

Problem B.6. Let $\gamma = 1/r_t$ and $\delta = 1/r_{t+1}$. Substitution into Problem B.1. and (B.5) yields

$$\frac{w_0 + w_1\delta}{1 + \delta} + \frac{w_0 + w_1\gamma}{1 + \gamma} = 1.$$

Some calculation provides $\gamma\delta = 1$.

Problem B.7. a) Introducing notation $r_t = \xi$, golden rule 2-cycle is equivalent to $r_{t+1} = 1/\xi$. Then (B.5) and the solution of Problem B.2 simplify to

$$\frac{w_0\Phi\xi^\mu - w_1\xi}{1 + \Phi\xi^\mu} = \frac{w_0\Phi\xi^{1-\mu} - w_1}{1 + \Phi\xi^{-\mu}}.$$

After some calculations, we end up with a nonalgebraic equation:

$$(w_1 + \Phi^2 w_0)\xi - \Phi\xi^\mu + \Phi\xi^{1-\mu} - (w_1 + \Phi^2 w_0) = 0.$$

It is obvious that one real root is equal to 1, and the others (if they exist), are the components of the 2-cycle.

b) For $\mu = 2/3$, ξ and ξ^μ are the simple powers of $\xi^{1-\mu}$. Therefore introducing $\chi = \xi^3$, now the generally transcendent equation reduces to a cubic equation. Dividing by $\chi - 1$ yields the quadratic equation

$$(w_1 + \Phi^2 w_0)\chi^2 + (w_1 + \Phi^2 w_0 - \Phi)\chi + (w_1 + \Phi^2 w_0) = 0$$

which can be solved even on the back of an envelope. For $w_1 = 1$, a positive solution exists if and only if $\beta < 1/27$. $\beta = 1/30$ yields $r_0 = 0.376$ which is not very interesting, since the expression $S(r)$ changes only by 0.003 in the whole interval $0.2 \le r \le 1$.

Appendix C

Problem C.1. We have the ith Lagrange function with U_i and multiplier ξ_i:

$$L_i\left({}_t c_{i,t}, \ldots, {}_t c_{D,t+D-i}\right) = \sigma^{-1} \sum_{i=0}^{D-i} \beta^j \, {}_t c_{i+j,t+j}^{\sigma}$$

$$+ \xi_i \left(r_t a_{i-1,t-1} + \sum_{i=0}^{D-i} (w_{i+j} - {}_t c_{i+j,t+j}) \, {}_t R_{t,t+j}^{-1} \right).$$

Take its partial derivative with respect to ${}_t c_{i+j,t+j}$ and set it to zero. With manipulation:

$$_t c_{i+j,t+j}^{\sigma-1} = \xi_i \beta^{-j} \, {}_t R_{t,t+j}^{-1}.$$

Taking into account that $\sigma = -\mu/(1 - \mu)$, we obtain

$$_t c_{i+j,t+j} = \Phi^j \xi_i^{\mu-1} \, {}_t R_{t,t+j}^{1-\mu}.$$

Substituting the last equation into (C.9), yields

$$\xi^{\mu-1} = \frac{r_t a_{i-1,t-1} + \sum_{j=0}^{D-i} w_{i+j} R_{t,t+j}^{-1}}{\sum_{j=0}^{D-i} \Phi^j \, {}_t R_{t,t+j}^{-\mu}}.$$

Using notations (C.15)–(C.16), implies

$$_t c_{i+j,t+j} = \Phi^j \, {}_t R_{t,t+j}^{1-\mu} \frac{r_t a_{i-1,t-1} + W_{i,t}}{V_{i,t}}.$$

For $j = 0$, this equation reduces to (C.17). ∎

Problem C.2. Programming.

Problem C.3. Plug in $w_j = 1/D, j = 0, \ldots, D$, $r^* = 1/\beta$ into (C.15°)–(C.16°) and (C.21°): $W(1/\beta) = (\sum_{i=0}^{D} r^{-i})/D$, $V(1/\beta) = (\sum_{i=0}^{D} r^{-i})/D$, $H(1/\beta) = 1$, $c_j(1/\beta) = 1/(D+1)$.

Problem C.4. Following Example B.7, $W = (1 + r^{-1} + r^{-2})/3$, $V = 3$, $H = (1 + r^{-1} + r^{-2})/9$, $W = (1 + r^{-1})/3$. $v = SV - W = 9 - r - 1 - r^{-1} - r^2 - r - 1$, and so on.

REFERENCES

AARON, H. J. (1966) "The Social Insurance Paradox", *Canadian Journal of Economics and Political Science 32* 371–4.

AIYAGARI, S. R. (1988) "Nonmonetary Steady States in Stationary Overlapping Generations Models with Long Lived Agents and Discounting: Multiplicity, Optimality, and Consumption Smoothing", *Journal of Economic Theory 45* 102–27.

AIYAGARI, S. R. (1989) "Can there be Short-Period Deterministic Cycles when People are Long-Lived?" *Quarterly Journal of Economics 104* 163–85.

ANDERSON, P. W., ARROW, K. J. and PINES, D., eds. (1988) *The Economy as an Evolving Complex System*, Redwood City, CA, Addison-Wesley.

ANDO, A. and MODIGLIANI, F. (1963) "The 'Life Cycle' Hypothesis of Saving: Aggregate Implications and Tests", *American Economic Review 53* 55–84.

AOKI, M. (1976) *Optimal Control and System Theory in Dynamic Economic Analysis*, New York, North Holland.

ARNOLD, V. I. (1973) *Ordinary Differential Equations*, Cambridge, MA, MIT Press. Second Russian edition: 1984.

ARROW, K. J. (1968) Application of Control Theory to Economic Growth, *Lectures in Applied Mathematics, Mathematics of Decision Sciences, Part 2, Vol. 12*, Providence RI, AMS.

ARROW, K. (1970) *Essays in the Theory of Risk Bearing*, Chicago, Markham.

ARROW, K. J., BLOCK, H. D. and HURWICZ, L. (1959) "On the Stability of the Competitive Equilibrium: II", *Econometrica 27* 82–109.

ARROW, K. J. and DEBREU, G. (1954) "Existence of Equilibrium for a Competitive Economy", *Econometrica 22* 265–90.

ARROW, K. J. and HAHN, F. (1971) *General Competitive Analysis*, San Francisco, Holden-Day.

ARROW, K. J. and HONKAPOHJA, S. (1985a) "Introduction", *Arrow and Honkapohja, eds.* 1–27.

ARROW, K. J. and HONKAPOHJA, S., eds. (1985b) *Frontiers of Economics*, Oxford, Blackwell.

ARROW, K. J. and HURWICZ, L. (1958) "On the Stability of the Competitive Equilibrium: I", *Econometrica 26* 522–52.

ARROW, K. J. and INTRILLIGATOR, M. D., eds. (1981) *Handbook of Mathematical Economics, Vol. I.* Amsterdam, North-Holland.

ARTHUR, W. B. (1989) "Competing Technologies, Increasing Returns, and Lock-in by Historical Events", *Economic Journal 99* 116–31.

ARTHUR, W. B. and MCNICOLL, G. (1978) "Samuelson, Population and Intergenerational Transfers", *International Economic Review 19* 241–6.

ATHANS, M. (1972) "The Discrete Time, Linear-Quadratic-Gaussian Stochastic Control Problem", *Annals of Economic and Social Measurements 1* 449–92.

ATHANS, M. (1975) "Theory and Application: A Survey of Decentralized Control Methods", *Annals of Economic and Social Measurements 4* 345–55.

ATKINSON, A. B. (1969) "The Timescale of the Economic Models: How Long is the Long Run", *Review of Economic Studies 36* 137–52.

AUERBACH, A. J. and KOTLIKOFF, L. J. (1987) *Dynamic Fiscal Policy*, Cambridge, Cambridge University Press.

AUGUSZTINOVICS, M. (1989) "The Costs of Human Life", *Economic Systems Research 1* 5–26.

AUGUSZTINOVICS, M. (1992) "Towards a Theory of Stationary Populations", *manuscript,* Institute of Economics, Budapest (earlier version: Discussion Paper, 1991).

AZARIADIS, C. (1993) *Intertemporal Macroeconomics*, Oxford, Blackwell.

BALASKO, Y., CASS, D. and SHELL, K. (1980) "Existence of Competitive Equilibrium in a General Overlapping-Generations Model", *Journal of Economic Theory 23* 307–22.

BALASKO, Y. and GHIGLINO, C. (1995) "On the Existence of Endogenous Cycles", *Journal of Economic Theory* 67 566–77.

BANAI, M. and LUKÁCS, B. (1987) "Investment Path and Variational Methods", *Közgazdasági Szemle* 34 432–40 (in Hungarian).

BAUER, T. (1978) "Investment Cycles in Centrally Planned Economies", *Acta Oeconomica* 21 243–60.

BAUMOL, W. J. (1970) *Economic Dynamics: An Introduction*, New York, MacMillan (third edition).

BELLMAN, R. E. (1957) *Dynamic Programming*, Princeton, Princeton University Press.

BENASSY, J.-P. (1974) "Neo-Keynesian Disequilibrium Theory in a Monetary Economy", *Review of Economic Studies* 41 87–104.

BENHABIB, J. (1992) *Cycles and Chaos in Equilibrium*, Princeton, Princeton University Press.

BENHABIB, J. and DAY, R. H. (1982) "A Characterization of Erratic Dynamics in the Overlapping Generations Model", *Journal of Economic Dynamics and Control* 4 37–55.

BERMAN, A. and PLEMONS, R. J. (1979) *Nonnegative Matrices in Mathematical Sciences*, New York, Academic Press.

BLACKWELL, D. (1965) "Discounted Dynamic Programming", *Annals of Mathematical Statistics* 36 226–35.

BLANCHARD, O. J. and FISCHER, S. (1989) *Lectures on Macroeconomics*, Cambridge, MA, MIT Press.

BLATT, J. M. (1978) "On the Econometric Approach to Business Cycle Modelling", *Oxford Economic Papers* 30 292–300.

BLATT, J. M. (1980) "On the Frisch Model of the Business Cycle", *Oxford Economic Papers* 32 467–79.

BLATT, J. M. (1983) *Dynamic Economic Systems*, Armonk N.Y., M. E. Sharpe.

BODEWIG, E. (1959) *Matrix Calculus*, Amsterdam, North Holland.

BOLDRIN, M. and MONTRUCCHIO, L. (1986) "On the Indeterminacy of Capital Accumulation Paths", *Journal of Economic Theory* 40 26–39.

BOLDRIN, M. and WOODFORD, M. (1990) "Equilibrium Models Displaying Endogenous Fluctuations and Chaos: A Survey", *Journal of Monetary Economics* 25 189–222.

BOYER, C. B. (1968) *A History of Mathematics*, Princeton, Princeton University Press.

BROCK, W. A. (1986) "Distinguishing Random and Deterministic

Systems, Abridged Version", *Journal of Economic Theory 40* 168–95.

BROCK, W. A. and HOMMES, C. H. (1997) "A Rational Route to Randomness", *Econometrica 65* 1059–95.

BRÓDY, A. (1970) *Planning, Proportions and Prices*, Amsterdam, North Holland.

BRÓDY, A. (1981) "On Control Models", *Kornai and Martos, eds* 149–61 (Hungarian original: 1973).

BRUNNER, K. and MELTZER, A., eds. (1976) *The Phillips Curve and Labor Markets*, Carnegie–Rochester Conference Series, Vol. 1, Amsterdam, North-Holland.

BRYSON, A. E. and HO, Y.-C. (1975) *Applied Optimal Control*, Washington D.C., Hemisphere Publishing C.

BURMEISTER, E. (1980) *Capital Theory and Dynamics*, Cambridge, Cambridge University Press.

CASS, D. (1965) "Optimum Growth in an Aggregate Model of Capital Accumulation", *Review of Economic Studies 32* 233–40.

CASS, D. (1966) "Optimum Growth in an Aggregative Model of Capital Accumulation: A Turnpike Theorem", *Econometrica 34* 833–50.

CHAMPSOUR, P. et al., ed. (1990) *Essays in Honor of Edmund Malinvaud*, Cambridge, MA, MIT Press.

CHIANG, A. (1984) *Fundamental Methods of Mathematical Economics*, New York, McGraw Hill.

CHIANG, A. (1992) *Elements of Dynamic Optimization*, New York, McGraw Hill.

CHIKÁN, A. (1984a) "Inventory Fluctuations (Cycles in the Hungarian Economy)", *Chikán, ed.*

CHIKÁN, A., ed. (1984b) *New Results in Inventory Research*, Amsterdam–Budapest, Elsevier, North-Holland, Akadémia.

CHOW, G. C. (1975) *Analysis and Control of Dynamic Economic Systems*, New York, Wiley.

CHOW, G. C. (1989) "Rational and Adaptive Expectations in Present Value Models", *Review of Economics and Statistics 71* 376–84.

CODDINGTON, E. A. and LEVINSON, N. (1955) *The Theory of Ordinary Differential Equations*, New York, McGraw Hill.

CUGNO, F. and MONTRUCCHIO, L. (1984) "Some New Techniques for Modelling Nonlinear Economic Fluctuations: A Brief Survey", *Goodwin et al., eds.* 146–65.

DAY, R. (1982) "Irregular Growth Cycles", *American Economic Journal* 72 406–14.

DAY, R. (1994) *Complex Economic Dynamics*, Cambridge, MA, MIT Press.

DAY, R. and PIAGINIANI, G. (1991) "Statistical Dynamics and Economics", *Journal of Economic Behavior and Organization 16* 37–83.

DEATON, A. (1992) *Understanding Consumption*, Oxford, Clarendon Press.

DEBREU, G. (1974) "Excess Demand Functions", *Journal of Mathematical Economics 1* 15–22.

DECHERT, W. D. (1984) "Does Optimal Growth Precludes Chaos? A Theorem on Monotonicity", *Journal of Economics 44* 57–61.

DENECKERE, R. and PELIKAN, S. (1986) "Competitive Chaos", *Journal of Economic Theory 40* 13–25.

DEVANEY, R. L. (1989) *An Introduction to Chaotic Dynamic Systems*, Redwood City, Addison-Wesley Publishing Company, Second edition.

DIAMOND, P. A. (1965) "National Debt in a Neoclassical Growth Model", *American Economic Review 55* 1126–50.

DOMAR, E. E. (1946) "Capital Expansion, Rate of Expansion and Employment", *Econometrica 14* 137–47.

DOMAR, E. E. (1957) *Essays in the Theory of Economic Growth*, New York, Oxford University Press.

ELAGDI, S.N. (1991) *An Introduction to Difference Equations*, New York, Springer.

ELBERS, C. and WEDDEPOHL, H. N. (1986) "Steady State Equilibria with Saving for Retirement in a Continuous Time Overlapping Generations Model", *Journal of Economics 46* 253–82.

EZEKIEL, M. (1938) "The Cobweb Theorem", *Quarterly Journal of Economics 52* 255–80.

FEINSTEIN, G. H., ed. (1967) *Socialism, Capitalism, and Economic Growth*, Cambridge University Press, Cambridge.

FELLNER, W. et al. (1967) *Ten Economic Studies in the Tradition of Irving Fisher*, New York, Wiley.

FRISCH, R. (1933) "Propagation Problems and Impulse Problems in Dynamic Economics", *Economic Essays in Honor of Gustav Cassel*, London, 171–205.

FULLER, A. and FISHER, M. (1958) "On the Stabilization of Matrices and the Convergence of Linear Iterative Processes", *Proceed-*

ings of the Cambridge Philosophical Society 54 417–25.

GALE, D. (1963) "A Note on the Global Instability of the Competitive Equilibrium", *Naval Research Quarterly 10* 81–9.

GALE, D. (1973) "Pure Exchange Equilibrium of Dynamic Economic Models", *Journal of Economic Theory 6* 12–36.

GALE, D. (1974) "The Trade Imbalance Story", *Journal of International Economics 4* 119–37.

GANDOLFO, G. (1971) *Mathematical Methods and Models in Economic Dynamics*, Amsterdam, North Holland. Second edition, 1983, third edition, 1995.

GANTMACHER, F. R. (1959) *The Theory of Matrices, Volumes 1 and 2*, New York, Chelsea.

GELFAND, I. M. and FOMIN, S. V. (1962) *Calculus of Variations*, Englewood Cliff, N.J., Prentice Hall.

GHIGLINO, C. and TVEDE, M. (1995) "No-Trade and the Uniqueness of Steady States", *Journal of Economic Dynamics and Control 19* 655–61.

GOKHALE, J., KOTLIKOFF, L. and SABELHAUS, J. (1996) "Understanding the Postwar Decline in U.S. Saving: A Cohort Analysis", *Brookings Papers on Economic Acticity (1)* 315–407.

GOODWIN, R. M. (1951) "The Nonlinear Accelerator and the Persistence of Business Cycles", *Econometrica 19* 1–17.

GOODWIN, R. M. (1967) "A Growth Cycle", *Feinstein, ed.* 54–8.

GOODWIN, R. M., KRÜGER, M. and VERCELLI, A., eds. (1984) *Nonlinear Models of Fluctuating Growth. Lecture Notes in Economics and Mathematical Systems 248* 146–65, Berlin, Springer.

GRANDMONT, J.-M. (1985) "On Endogenous Business Cycles", *Econometrica 53* 995–1045.

GRANDMONT, J.-M. (1986) "Periodic and Aperiodic Behavior in Discrete One-Dimensional Dynamic Systems", *Hildenbrand and Mas Collel, eds.* 227–65.

GRANDMONT, J.-M. (1998) "Expectations Formation and Stability of the Large Socioeconomic Systems", *Econometrica 66* 741–81.

GRANDMONT, J.-M. and LAROQUE, G. (1990) "Stability, Expectations and Predetermined Variables", *Champsour et al., eds.* VOL. 1. 71–92.

GUCKENHEIMER, J. (1979) "Sensitive Dependence to Initial Conditions for One-Dimensional Maps", *Communications of Mathematical Physics 70* 133–60.

GUCKENHEIMER, J. and HOLMES, P. (1986) *Nonlinear Oscillations, Dynamic Systems, and Bifurcations of Vector Fields*, Second enlarged and revised edition, New York, Springer.

HAHN, W. (1967) *Stability of Motion*, New York, Springer.

HALANAY, A. and SAMUEL, J. (1997) *Differential Equations, Discrete Systems and Control, Economic Models*, Dordrecht, Kluwer.

HALMOS, P. (1958) *Finite-Dimensional Vector Spaces*, New York, Springer.

HARROD, R. (1939) "An Essay on Dynamic Economics", *Economic Journal 49* 14–33.

HARROD, R. (1948) *Towards a Dynamic Economics*, London, Macmillan.

HAYEK, F. A. (1935) *Collectivistic Economic Planning*, London, Routledge and Kegan Paul.

HICKS, J. (1950) *A Contribution to the Theory of Trade Cycle*, Oxford, Clarendon.

HILDENBRAND, W. and MAS COLLEL, A., eds. (1986) *Contributions to Mathematical Economics*, Amsterdam, Elsevier.

HILDENBRAND, W. and SONNENSCHEIN, H., eds. (1991) *Handbook of Mathematical Economics Vol. IV*, Amsterdam, North-Holland.

HOLLY, S., RÜSTEM, B. and ZARROP, M. B., eds. (1979) *Optimal Control for Econometric Models (An Approach to Economic Policy Formulation)*, London, Macmillan.

HOMMES, C. H. (1991) *Chaotic Dynamics in Economic Models: Some Simple Case-Studies*, Groningen Theses in Economics, Management and Organization, Wolters-Nordhoff.

HOMMES, C. H. (1993) "Periodic, Almost-Periodic and Chaotic Dynamics in Hicks' Nonlinear Trade Cycle Model", *Economic Letters 41* 391–7.

HOMMES, C. H. (1994) "Dynamics of the Cobweb Model with Adaptive Expectations and Nonlinar Supply and Demand", *Journal of Economic Behavior and Organization 24* 315–35.

HOMMES, C. H. and NUSSE, H. E. (1989) "Does an Unstable Keynesian Unemployment Equilibrium in a Non-Walrasian Dynamic Macroeconomic Model Imply Chaos?", *Scandinavian Journal of Economics 91* 161–7.

HOMMES, C. H. and NUSSE, H. E. (1992) "Period Three to Period Two Bifurcation for Piecewise Linear Models", *Journal of Economics 54* 157-69.

HOMMES, C. H., NUSSE, H. E. and SIMONOVITS, A. (1995) "Cycles and Chaos in a Socialist Economy" *Journal of Economic Dynamics and Control 19* 155–79.

HOMMES, C. H. and SORGER, G. (1997) "Consistent Expectations Equilibria", *Macroeconomic Dynamics 2* 287-321.

HONKAPOHJA, S. and ITO, T. (1980) "Inventory Dynamics in a Simple Disequilibrium Macroeconomic Model", *Scandinavian Journal of Economics 82* 184–98.

ICKES, B. W. (1986) "Cyclical Fluctuations in Centrally Planned Economies", *Soviet Studies 38* 36–52.

ITO, S., TANAKA, S. and NAKADA, H. (1979) "On Unimodal Linear Transformations and Chaos: II", *Tokyo Journal of Mathematics 2* 241–59.

JAKOBSON, M.V. (1981) "Absolutely Continuous Invariant Measures for One-Parameter Families of One-Dimensional Maps", *Communications of Mathematical Physics 81* 39–81.

KALMAN, R. (1960) "A Contribution to the Theory of Optimal Control", *Bol. Socied. Mat. Mexicana 5* 102–19.

KAMIEN, M. I. and SCHWARTZ, N. L. (1981) *Dynamic Optimization: The Calculus of Variations and Optimal Control in Economics and Management*, Amsterdam, North-Holland. Second edition, 1991.

KEHOE, T. J. (1991) "Computation and Multiplicity of Equilibria", *Hildenbrand and Sonnenschein, eds.*

KEHOE, T. J. and LEVINE, D. K. (1984) "Regularity and Overlapping Generations Exchange Economies", *Journal of Mathematical Economics 13* 69-93.

KEHOE, T. J. and LEVINE, D. K. (1985) "Comparative Static and Perfect Foresight in Infinite Horizon Economics", *Econometrica 53* 433–53.

KEHOE, T. J. and LEVINE, D. K. (1990) "The Economics of Indeterminacy in Overlapping Generations Models" *Journal of Public Economics 42* 219–43.

KENDRICK, D. (1981) "Control Theory with Applications to Economics", *Arrow and Intrilligator, eds.* 111–58.

KEYNES, J. M. (1936) *The General Theory of Employment, Interest and Money*, New York, Harcourt, Brace.

KIM, O. (1983) "Balanced Equilibrium in a Consumption Loans Model", *Journal of Economic Theory 29*, 339–46.

KIRMAN, A. (1992) "Whom or What Does the Representative Individual Represent?" *Journal of Economic Perspectives 6* 117–36.

KOOPMANS, T. C. (1965) "On the Concept of Optimal Economic Growth", in *Semain d'Etude sur le Role de l'Analyse Econometrique dans la Formulation due Plans de Dévelopment*, Vatican City, Pontifical Academy of Sciences, Vol. I. 225–87.

KOOPMANS, T. C. (1967) "Objectives, Constraints and Outcomes in Optimal Growth Models", *Econometrica 35* 1–15.

KORNAI, J. (1971) *Anti-Equilibrium*, Amsterdam, North-Holland.

KORNAI, J. (1980) *The Economics of Shortage*, Amsterdam, North-Holland.

KORNAI, J. (1982) *Growth, Shortage and Efficiency*, Oxford: Basil Blackwell.

KORNAI, J. (1992) *The Socialist System. The Political Economy of Communism*, Oxford University Press.

KORNAI, J. and MARTOS, B. (1973) "Autonomous Functioning of the Economic System", *Econometrica 41* 509–28.

KORNAI, J. and MARTOS, B. (1981a) "Introduction: Theoretical Background of the Research", *Kornai and Martos, eds.* 17–56.

KORNAI, J. and MARTOS, B., eds. (1981b) *Non-Price Control*, Amsterdam, North-Holland.

KORNAI, J. and SIMONOVITS, A. (1977) "Decentralized Control Problems in Neumann-economies", *Journal of Economic Theory 14* 44–67.

KORNAI, J. and SIMONOVITS, A. (1981) "Stock-Signal Model Regulated from a Normal Path", *Kornai and Martos, eds.* 223–45.

KOVÁCS, J. and VIRÁG, I. (1981) "Periodic versus Continuous Growth" *Acta Oeconomica 27* 41–55.

KUHN, H. W. and SZEGŐ, G. eds. (1969) *Mathematical Systems Theory and Economics, I.* Berlin, Springer.

KURIHARA, K. K., ed. (1954) *Post-Keynesian Economics*, New Brunswick, Rutgers University Press.

KYDLAND, F. E. and PRESCOTT, E. C. (1977) "Rules rather than Discretion: The Inconsistency of Optimal Plans", *Journal of Political Economics 85* 473–91.

LACKÓ, M. (1980) "Cumulation and Easing of Tensions", *Acta Oeconomica 24* 357-78.

LAITNER, J. P. (1981) "The Stability of Steady States in Perfect Foresight Models", *Econometrica 49* 319–33.

LAITNER, J. P. (1984) "Transition Time Paths for Overlapping-Generations Models", *Journal of Economic Dynamics and Control* 7 111–29.

LANCASTER, K. (1968) *Mathematical Economics*, New York, Macmillan.

LANCASTER, P. (1969) *Theory of Matrices*. New York, Academic Press.

LEONTIEF, W. W. (1941) *The Structure of the American Economy*, New York, Oxford University Press (edition in 1951).

LEUNG, S. F. (1994) "Uncertain Lifetime, The Theory of Consumer, and the Life Cycle Hypothesis", *Econometrica* 62 1233–9.

LEVHARI, D. and MIRMAN, L. J. (1980) "The Great Fish War: An Example Using a Dynamic Cournot-Nash Solution", *The Bell Journal of Economics* 11 322–34.

LI, J. A. and YORKE, J. A. (1975) "Period Three Implies Chaos", *American Mathematical Monthly* 82 985–92.

LORENZ, E. N. (1963) "Deterministic Nonperiodic Flow", *Journal of Atmospheric Sciences* 20 130–41.

LORENZ, E. N. (1993) *The Essence of Chaos*, Seattle, University of Washington Press.

LORENZ, H.-W. (1993) *Nonlinear Dynamics Economics and Chaotic Motion*, Lecture Notes in Economics and Mathematical Systems, Berlin, Springer, second edition.

LOVELL, M. C. (1962) "Buffer Stocks, Sales Expectations and Stability: A Multi-Sector Analysis of the Inventory Cycle", *Econometrica* 30 267–96.

LOVELL, M. C. (1986) "Tests of Rational Expectations Hypotheses", *American Economic Review* 76 110–24.

LUCAS, R. E. (1976) "Econometric Policy Evaluation, A Critique", *Brunner and Meltzer, eds.* 19–46.

LUCAS, R. E. (1987) *Models of Business Cycles*, Oxford, Basil Blackwell.

LYAPUNOV, A. (1893) *Stability of Motions*, New York, Academic Press (English translation from the Russian original) 1966.

MANUELLI, R. E. and SARGENT, T. J. (1987) *Exercises in Dynamic Macroeconomic Theory*, Cambridge MA, Harvard University Press.

MARTOS, B. (1981) "Concepts and Theorems from Control Theory", *Kornai and Martos, eds.* 81–112.

MARTOS, B. (1990) *Economic Control Structures*, Amsterdam, North Holland.

McFADDEN, D. (1969) "On the Controllability of Decentralized Macroeconomic Systems, The Assignment Problem", *Kuhn and Szegő, eds.* 221–39.

McKENZIE, L. W. (1986) "Optimal Economic Growth, Turnpike Theorems and Comparative Dynamics", *Arrow and Intrilligator, eds.* 1281–355.

MEDIO, A. (1991) "Continuous-Time Models of Chaos in Economics", *Journal of Economic Behavior and Organization 16* 115–51.

MEDIO, A. (1992) *Chaotic Dynamics, Theory and Applications to Economics*, Cambridge, Cambridge University Press.

MEDIO, A. (1995) "Ergodicity, Predictability and Chaos", *Discussion Paper*, University of Venice.

METZLER, L. (1941) "The Nature and Stability of Inventory Cycles", *Review of Economic Statistics 23* 113–29.

MITRA, T. (1998) "On the Relationship between Discounting and Complicated Behavior in Dynamic Optimization Models", *Journal of Economic Behavior and Organization*.

MITRA, T. and SORGER, G. (1999): "Rationalizing Policy Functions by Dynamic Optimization", *Econometrica 67* 375–92.

METZLER, L. (1945) "The Stability of Multiple Markets: The Hicks Conditions", *Econometrica 13* 277–92.

MODIGLIANI, F. and BRUMBERG, R. (1954) "Utility Analysis and the Consumption Function: An Interpretation of Cross-Section Data", *Kurihara, ed.* 388–436.

MOLNÁR, GY. and SIMONOVITS, A. (1996) "Expectations, (In)-stability and (In)viability in Realistic Overlapping Cohorts Models", *Discussion Paper 40*, Institute of Economics, Budapest.

MOLNÁR, GY. and SIMONOVITS, A. (1998) "Expectations, (In)-stability and (In)viability in Realistic Overlapping Cohorts Models", *Journal of Economic Dynamics and Control 23* 303–32.

MORISHIMA, M. (1964) *Growth, Stability and Equilibrium*, Oxford, Oxford University Press.

NELSON, R. and WINTER (1982) *An Evolutionary Theory of Economic Change*, Cambridge MA, Belknap Press.

NEUMANN, J. (1938) "Model of General Equilibrium", *Review of Economic Studies 13* (1945) 1–18. (Translated from German original).

PETERS, W. (1988) "A Pension Insurance System in an Overlapping Generations Model", *Journal of Institutional and Theoretical Economics 144* 813–30.

PHELPS, E. S. (1961) "The Golden Rule of Accumulation: A Fable for Growthmen" *American Economic Review 51* 638–43.

PHILLIPS, W. (1954) "Stabilization Policy and a Closed Economy", *Economic Journal 64* 290–323.

PITCHFORD, J. D. and TURNOVSKY, S., eds. (1977) *Application of Control Theory to Economic Analysis*, Amsterdam, North-Holland.

POHJOLA, M. T. (1981) "Stable and Chaotic Growth: the Dynamics of a Discrete Version Cycle Model", *Journal of Economics 41* 27–39.

PÓLYA, GY. (1968) *Mathematics and Plausible Reasoning, Vol. I: Induction and Analogy*, second edition, Princeton, University Press.

PÓLYA, GY. and SZEGŐ, G. (1924) *Problems and Theorems in Analysis, Vol. II*, Berlin, Springer, translation from German, 1976.

PONTRYAGIN, L. S. (1962) *Ordinary Differential Equations*, Reading, MA, Addison Wesley.

PONTRYAGIN, L. S., BOLTYANSKII, V. G., GAMKRELIDZE, R. V. and MISHCHENKO, J. F. (1962) *The Mathematical Theory of Optimal Processes*, New York, Wiley.

PRESTON, A. J. (1977) "Existence, Uniqueness and Stability of Linear Optimal Stabilization", *Pitchford and Turnovsky, eds.* 293–335.

RALSTON, A. (1965) *A First Course in Numerical Analysis*, New York, McGraw Hill.

RAMSEY, F. (1928) "A Mathematical Theory of Savings", *Economic Journal 38* 543–59.

REICHLIN, P. (1992) "Endogenous Cycles with Long-Lived Agents", *Journal of Economic Dynamics and Control 16* 243–66.

RUDIN, W. (1976) *Principles of Mathematical Analysis*, New York, McGraw Hill, Third edition.

SAARI, D. G. (1985) "Iterative Price Mechanisms", *Econometrica 53* 1117–31.

SALMON, M. and YOUNG, P. (1979) "Control Models and Quantitative Economic Policy", *Holly et al., eds.* 74–105.

SAMUELSON, P. A. (1939a) "Interactions between the Multiplier Analysis and the Principle of Acceleration", *Review of Economic Studies 21* 75–8.

SAMUELSON, P. A. (1939b) "A Synthesis of the Principle of Acceleration and the Multiplier", *Journal of Political Economy 47* 786–97.

SAMUELSON P. A. (1941) "The Stability of Equilibrium: Comparative Statistics and Dynamics",*Econometrica 9* 97–120.

SAMUELSON, P. A. (1947) *Foundations of Economics Analysis*, Cambridge, MA, Harvard University Press, enlarged edition, 1983.

SAMUELSON, P. A. (1958) "An Exact Consumption-Loan Model of Interest with or without the Social Contrivance of Money", *Journal of Political Economy 66* 467–82.

SAMUELSON, P. A. (1965) "A Catenary Turnpike Theorem Involving Consumption and the Golden Rule", *American Economic Review 55* 486–96.

SAMUELSON, P. A. (1975) "The Optimum Growth Rate for Population", *International Economic Review 16* 531–37.

SARGENT, T. J. (1987) *Dynamic Macroeconomic Theory*, Cambridge MA, Harvard University Press.

SCARF, H. (1960) "Some Examples of Global Instability of Competitive Equilibrium", *International Economic Review 1* 157–72.

SCARF, H. (1967) "On the Approximation of Fixed Points of a Continuous Mapping", *SIAM Journal of Applied Mathematics 15* 1328 –43.

SEN, A., ed. (1970) *Growth Economics*, Harmondsworth, Penguin.

SHARKOVSKII, A. N. (1964) "Coexistence of Cycles of a Continuous Map of the Line into Itself", *Ukranian Mathematical Journal 16* 51–71 (in Russian), English translation, Stefan (1977).

SHELL, K. (1967a) "Optimal Programs of Capital Accumulation for an Economy in which there is Exogenous Technical Change", *Shell, ed.* 1–30.

SHELL, K., ed. (1967b) *Essays on the Theory of Optimal Economic Growth*, Cambridge, MA, MIT Press.

SHILLER, R. (1978) "Rational Expectations and the Dynamic Structure of Macroeconomic Models", *Journal of Monetary Economics 4* 1–44.

SIMON, H. A. (1956) "Dynamic Programming under Uncertainty with a Quadratic Criterion Function", *Econometrica 24* 74–81.

SIMONOVITS, A. (1981a) "Constrained Control and Destabilization", *Kornai and Martos, eds.* 281–320.

SIMONOVITS, A. (1981b) "Maximal Convergence Speed of Decentralized Control", *Journal of Economic Dynamics and Control 3*

51–64.

SIMONOVITS, A. (1982) "Buffer Stocks and Naive Expectations in a Non-Walrasian Dynamic Macrodynamic Model: Stability, Cyclicity and Chaos", *Scandinavian Journal of Economics 84* 571–81.

SIMONOVITS, A. (1991) "Investment Limit Cycles in the Socialist Economy", *Economics of Planning 24* 27–46.

SIMONOVITS, A. (1992) "Indexed Mortgages and Expectations: Mathematical Analysis and Simulation", *Acta Oeconomica 44* 144–60.

SIMONOVITS, A. (1995a) "On the Number of Balanced Steady States in Overlapping Cohorts Model", *Acta Oeconomica 47* 51–67.

SIMONOVITS, A. (1995b) "Once More on Optimal Growth", *Közgazdasági Szemle 42* 1136–46 (in Hungarian).

SIMONOVITS, A. (1999a) "Linear Decentralized Control Model with Expectations", *Economic Systems Research*.

SIMONOVITS, A. (1999b) "Are there Endogenous Cycles in a Realistic Overlapping Cohorts Model?", *Structural Change and Economic Dynamics 10* 261–75.

SIMONYI, K. (1981) *The Cultural History of Physics*, (in Hungarian) Budapest, Gondolat, Second, enlarged edition.

SIMS, C. A. (1986) "Comments", *Sonnenschein, ed.* 37–9.

SINGER, D. (1978) "Stable Orbits and Bifurcations of Maps of the Interval", *SIAM Journal of Applied Mathematics 35* 260–7.

SLUTZKY, E. (1927) "The Summation of Random Causes as the Sources of Cyclic Processes", *Econometrica 5* 105–46 (revised translation from Russian) 1937.

SOLOW, R. (1956) "A Contribution to the Theory of Economic Growth", *Quarterly Journal of Economics 70* 65–94.

SOLOW, R. (1957) "Technical Change and the Aggregate Production Function", *Review of Economics and Statistics 39* 312–20.

SONNENSCHEIN, H. F., ed. (1986) *Models of Economic Dynamics*, New York, Springer.

SORGER, G. (1992) "On the Minimum Rate of Impatiance for Complicated Optimum Growth Paths", *Journal of Economic Theory 12* 11–30.

STEFAN. P. (1977) "A Theorem of Sharkovskii on the Existence of Periodic Orbits of Continuous Endomorphism of the Real Line", *Communications of Mathematical Physics 54* 237–48.

STOKEY, N. and LUCAS, R. (1989) *Recursive Methods in Economic Dynamics*, Cambridge MA, Harvard University Press (with a contribution: Prescott, E.)

STROTZ, R. H. (1955) "Myopia and Inconsistency on Dynamic Utility Maximization", *Review of Economic Studies 23* 165–80.

TAKAYAMA, A. (1974) *Mathematical Economics*, Hinsdale IL, Dryden, Second edition 1985.

THEIL, H. (1957) "A Note on the Certainty Equivalence in Dynamic Programming", *Econometrica 25* 346–9.

TINBERGEN, J. (1956) "The Optimum Rate of Savings", *Economic Journal 66* 603–9.

TINBERGEN, J. (1960) "Optimum Savings and Utility Maximization over Time", *Econometrica 28* 481–9.

TOBIN, J. (1967) "Life Cycle Saving and Balanced Growth", *Fellner et al., eds.* 231–56.

TSE, E. and ATHANS, M. (1972) "Adaptive Stochastic Control for a Class of Linear Systems", *IEEE Transaction on Automatic Control 17* 38–52.

VARGA, R. (1962) *Matrix Iterative Analysis*, Englewood Cliff, N.J., Prentice Hall.

VARIAN, H. (1981) "Control Theory with Economic Applications", *Arrow and Intrilligator, eds.* 111–58.

WALRAS, L. (1874, 1877) *Elements of Pure Economics*, London, Allen and Unwin (translated from the French original 1954).

WITSENHAUSEN, H. S. (1968) "A Counterexample in Stochatic Optimum Control", *SIAM Journal of Control 5* 131–47.

WONHAM, W. M. (1967) "Optimal Stationary Control of a Linear System with State-dependent Noise", *SIAM Journal of Control 5* 486–500.

YAARI, M. E. (1965) "Uncertain Lifetime, Life Insurance and the Theory of Consumer", *Review of Economic Studies 32* 137–50.

YOUNG, D. M. (1971) *The Iterative Solution of Large Linear Systems*, New York, Academic Press.

ZALAI, E. (1989) *Introduction to Mathematical Economics*, Budapest, KJK, (in Hungarian).

INDEX OF SUBJECTS